FLORA OF TROPICAL EAST AFRICA

ROSACEAE

(including Amygdalaceae and Chrysobalanaceae)

R. A. Graham

Herbs, shrubs, scramblers or trees. Leaves simple or compound, alternate, stipulate. Flowers regular or irregular, hermaphrodite, monoecious, dioecious, or polygamous, perigynous or epigynous. Calyx campanulate or urceolate, free or adnate in part to the receptacle, sometimes gibbous and asymmetrical, the lobes 4–5, with or without an additional row of 4–5 alternating outer lobes (epicalyx). Receptacle flat, or convex and sometimes very accrescent, or concave, the mouth symmetrically or asymmetrically crateriform. Petals 5 or 0. Stamens 1 to many, sometimes biseriate, exserted or included, free or basally connate, regularly disposed or inserted on one side of the receptacle only ; staminodes (if present) usually inserted on one side of the receptacle and there replacing the fertile stamens. Carpels 1 to many, free or fused to each other and to the receptacle, or fused but free of the receptacle. Styles terminal or basal, exserted or included. Ovules 1 or 2.

A world-wide family, especially numerous in temperate regions, containing many herbs or trees widely cultivated for amenity or food value.

The following key applies to tropical East African species only.

Epicalyx present ; herbs, often stoloniferous, or low shrubs :
  Flowers apetalous, 4-merous ; leaves simple, trifoliolate or palmately lobed ; receptacle slightly or not accrescent, remaining dry . . . . 1. **Alchemilla**
  Flowers with obvious white petals, 5-merous ; leaves trifoliolate only ; receptacle greatly accrescent, becoming fleshy and juicy . . . . . 2. **Fragaria**
Epicalyx absent ; trees, shrubs (some prickly), or scramblers :
  Prickly shrubs or scramblers . . . . . . 3. **Rubus**
  Trees or shrubs, without prickles :
    Leaves ericoid, linear or ± so, the margins usually revolute ; stigma with a feathery margin . 4. **Cliffortia**
    Leaves not as above ; stigma without a feathery margin :
      Leaves imparipinnate ; tall, dioecious or (rarely) polygamous tree . . . . . . 5. **Hagenia**
      Leaves simple :
        Style terminal . . . . 6. **Pygeum**
        Style basal :
          Fertile stamens inserted all round the receptacle-rim ; ovary inserted at the base of the calyx-tube . . . . 7. **Chrysobalanus**

Fertile stamens inserted only on the dorsal rim of the receptacle, with staminodes only inserted ventrally ; ovary inserted on the dorsal side of the calyx-tube :
Carpels 2-locular . . . . . . 8. **Parinari**
Carpels 1-locular :
Stamens and styles long-exserted ; staminodes very short, free to the base ; sepals and bracteoles often covered with stipitate glands . 9. **Hirtella**
Stamens and styles included ; staminodes basally united and disposed as a comb or elevated tongue ; sepals and bracteoles without stipitate glands . . . . . 10. **Magnistipula**

1. ALCHEMILLA

L., Sp. Pl. : 123 (1753) & Gen. Pl., ed. 5 : 58 (1754)

Perennial or annual (but not in tropical East Africa) herbs or low shrubs. Stems woody or herbaceous, erect or decumbent, often arising from a central rosette producing extensively creeping or scrambling non-rooting runners or stolons rooting here and there at the nodes. Leaves radical or cauline or both on the one plant, sessile to long-stalked, undivided or divided into a varying number of shallow or deep lobes arranged palmately or flabellately within a rounded to reniform outline ; petioles adnate in part or throughout to membranous or foliaceous stipules. Inflorescence simple or branched, racemose or paniculate, exceeding or ± hidden among the leaves, axillary, few- to many-flowered, sometimes congested. Flowers small, hermaphrodite, tetramerous, pedicellate to subsessile, subtended by a bract. Calyx-tube urceolate, ± membranous and persistent, bearing around the mouth two series each of 4 free lobes, the inner 4 (calyx-lobes*) alternating with the usually smaller outer 4 (epicalyx-lobes*) ; disc fleshy or somewhat so, closing the mouth of the calyx-tube. Petals 0. Stamens 4, inserted on the rim of the disc, alternate to the calyx-lobes. Carpels 1–12, stalked, inserted at the base of the receptacle ; styles basal ; stigma capitate. Achenes 1–12, included within the calyx-tube.

A large genus of temperate climates, and in tropical regions confined to the cooler conditions of mountains and highlands. Apomixis plays an important role in the life-cycle of temperate spp., but its occurrence is not definitely established in all those of tropical East Africa (see Hjelmquist in Bot. Not. 109: 21 (1956)).

1. Stems wholly woody, often largely or entirely hidden by closely adjacent stipules ; basal leaves absent 2
Stems herbaceous or ± woody in part, not hidden by stipules except at the extremities of young branches ; basal leaves produced or not . . . 12
2. Stipules apically entire, membranous throughout, usually reddish-brown 3
Stipules apically dentate, membranous throughout, or terminally ± foliaceous, or with a distinctly green and leafy upper part 10

* For brevity here and in the following key and descriptions these terms are used.

3. Leaves all undivided, or all lobed 4
   (Leaves both lobed and undivided on one plant . . . . 5 × 2. *A. stuhlmannii* × *subnivalis*)
4. Leaves all undivided (but usually terminally dentate) 5
   Leaves all 3–5-lobed 6
5. Leaves broadest at the apex, cuneate or ± so; terminal teeth up to 1·5(–2) mm. deep. . 1. *A. elgonensis*
   Leaves broadest below the apex, obovate; terminal teeth 0·5–1 mm. deep . . . . 2. *A. subnivalis*
6. Lobes 3 only, divided to the base or nearly so 7
   Lobes 3 or 5, not or not all divided to the base . . . 8
7. Lobes all about same size; lateral lobes terminally dentate . . . . 3. *A. triphylla*
   Lobes unequal, the lateral ones smaller (i.e. narrower) than the central one and terminally entire . . . . 4. *A. argyrophylla*
8. Lobes all of equal depth 9
   Lobes of unequal depth . . . . . . 7. *A. dewildemanii*
9. Leaves basally cuneate . . . . . . 5. *A. stuhlmannii*
   Leaves basally truncate or nearly so. . . . 6. *A. roccatii*
10. Stipules membranous throughout or terminally ± foliaceous though not or scarcely green; leaves basally cuneate to truncate, never cordate with a basal sinus 11
    Stipules distinctly green and leafy in the upper part; leaves basally cordate or ± so with a basal sinus (very rarely subtruncate), often markedly plicate . . . . . . 8. *A. johnstonii*
11. Stipules wholly membranous; leaves 3-lobed, basally cuneate or ± so; the lobes subequal, nearly reaching the base of the lamina . 3. *A. triphylla*
    Stipules membranous or terminally ± foliaceous; leaves 3–5-lobed, basally truncate or nearly so 6. *A. roccatii*
12. Calyx-tube glabrous 13
    Calyx-tube hairy 14
13. Tufts of basal leaves absent or at least not persistent; leaves 3–5-lobed; pedicels up to 1·5 mm. long; (if calyx-tubes glabrous except for 1 or 2 white hairs, and epicalyx-lobes clearly exceeding calyx-lobes, see 12, *A. cryptantha*) . . . . 9. *A. ellenbeckii*
    Tufts of basal leaves produced and persistent; leaves 7–9(–11)-lobed; pedicels 5–50 mm. long . . . . 13. *A. gracilipes*
14. Epicalyx-lobes exceeding the calyx-lobes 15
    Epicalyx-lobes shorter than, or sometimes about as long as but not exceeding the calyx-lobes. . . 17
15. Tufts of basal leaves present and persistent, 7-lobed; stolon-leaves 5–7-lobed, the lobes reaching beyond the centre of the lamina thus longer than broad; inflorescences of rather few-flowered axillary cymes, racemes, or panicles, rarely of 1–2 flowers hidden within the stipules on the stolons; achenes

5–8 ; (if the leaves with shallow lobes, not reaching the centre of the lamina and achenes not exceeding 2, see 14, *A. rothii*) . . 10. *A. kiwuensis*

Tufts of basal leaves absent or, at least not persistent, 5-lobed ; inflorescence of 1–2 flowers hidden within the stipules (in 12, *A. cryptantha* rarely in racemes, cymes or panicles) 16

16. Achenes 3–12 ; rooting stolons woody or herbaceous ; leaves small, the lobes sharply 5–7-toothed at the apex ; stipules on stolons entire, membranous and brown, or apically dentate with teeth becoming green (rarely ± foliaceous at the apex) . . . . . 11. *A. microbetula*

Achenes 5–8 ; rooting stolons herbaceous ; lobes of leaves 9–11-toothed at the apex ; stipules on rooting stolons or runners with a distinctly leafy upper half, or ± leafy throughout, not entire and membranous . . . . 12. *A. cryptantha*

17. Leaves basally truncate or very broadly cuneate, the basal lobes not enclosing a sinus . . 6. *A. roccatii*

Leaves basally ± cordate, the basal lobes enclosing a sinus 18

18. Flowers 1–2, hidden within the stipules . . 11. *A. microbetula*

Flowers not as above, but borne in cymes, racemes or panicles 19

19. Tufts of basal leaves produced 20

Tufts of basal leaves absent 23

20. Pedicels 3–50 mm. long, often filiform and obvious 21

Pedicels not or scarcely exceeding 1 mm. in length 22

21. Lobes of the leaves narrowest towards the apex, and usually reaching at least to the centre of the lamina and often beyond ; hairs of stems and petioles silky-white, appressed ; achenes usually 2–3 . . . . . . 13. *A. gracilipes*

Lobes of the leaves broadest towards the apex, seldom reaching to the centre of the lamina, often very shallow and the leaves appearing unlobed ; hairs of stems and petioles pale brown, spreading or a little deflexed ; achenes 1 (–2) . . . . 14. *A. rothii*

22. Plant stout, large ; radius of leaf-lamina (3–)4–7 cm., the lobes terminally rounded or obtuse . . . . 15. *A. fischeri*

Plant smaller ; radius of leaf-lamina 1·5–3 cm., the lobes terminally truncate or nearly so, sometimes emarginate . . . . . 16. *A. volkensii*

23. Leaves with 3–5(–7) lobes, not or scarcely exceeding 3·5 cm. in width and often much less ; stems usually woody . . . . 8. *A. johnstonii*

Leaves with (7–)9–17 lobes, 7–10 cm. in width ; stems stout but apparently not woody ; a large, handsome species . . . . 17. *A. hageniae*

1. **A. elgonensis** *Mildbr.* in N.B.G.B. 8 : 228 (1922) ; Hjelmq. in Bot. Not. 109 : 21 (1956) ; Hedb. in Symb. Bot. Ups. 15 (1) : 109 (1957). Type : Elgon, *Lindblom* (S, holo.)

A low spreading bush with woody branches, 15–120 cm. tall. Lateral branchlets 1·5–12 cm. long. Stipules ovate, entire, reddish-brown, transversely wrinkled, obscuring the branchlets, glabrous or with long white hairs on the margins. Leaves undivided, sessile or nearly so, simple, broadest at the apex and thence ± evenly narrowed to the base, 4–15 × 2–5 mm., apically truncate and ± tridentate, with the acute, forward-directed teeth up to 1·5(–2) mm. deep, covered on both sides with long silvery hairs. Flowers 1–3 together. Calyx-tube 1·5 mm. long ; calyx-lobes triangular, 1–1·5 mm. long ; epicalyx-lobes lanceolate, shorter than the calyx-lobes. Achenes usually 2.

UGANDA. Elgon, below Jackson's summit, *Liebenberg* 1616 !, *Saunders & Hancock* 69 !; Elgon, above Bulambuli, Sept. 1932, *A. S. Thomas* 559 !
KENYA. Elgon, Dec. 1930, *Lugard* 369 ! & Feb. 1932, *Tweedie* 24 ! ; Marakwet–Suk Hills, June 1935, *Dale* 3403 !
DISTR. **U**3 ; **K**3, ?5 ; not known elsewhere.
HAB. Upland moor and moist bamboo-thickets, in both wet and dry places, 2700–4350 m.

NOTE. *Dale* 3403 has a leaf somewhat intermediate between this species and the next, there being a tendency for maximum width below the apex in certain leaves. The flowers, however, correspond with those of *A. elgonensis* in size.

1 × 8. **A. elgonensis** *Mildbr.* × **johnstonii** *Oliv.*; Hedb. in Symb. Bot. Ups. 15 (1) : 116 (1957).

An example of what is possibly this hybrid has been collected in Kenya (from Elgon, May 1948, *Hedberg* 923 !) and differs from *A. elgonensis* in having subequally but shallowly trilobed leaves with broadly cuneate, almost truncate bases. There is considerable resemblance to some of the forms of *A. roccatii* (see p. 9) from which it differs in the leaves being 3-lobed only, and in their bases tending to be cuneate rather than truncate. The position with regard to hybrids of the suffruticose species is however not altogether clear due to paucity of material.

2. **A. subnivalis** *Bak.f.* in J.L.S. 38 : 250 (1908) ; Z.A.E. : 227, t. 22 E–H (1911) ; Hauman & Balle in Ann. Zool. Bot. Afr. 24 : 335 (1934) ; F.P.N.A. 1 : 253 (1948) ; F.C.B. 3 : 15 (1952) ; Hedb. in Symb. Bot. Ups. 15 (1) : 107 (1957). Type : Uganda, Ruwenzori, *Wollaston* (BM, holo. !)

A mat-forming, low-growing shrub up to 1 m. high, largely similar to the preceding but differing in leaf-shape. Main branches woody, up to 35 cm. long, red-brown, covered by stipules. Lateral branchlets short, 4–5(–10) cm. long. Stipules ± ovate, red-brown, membranous, transversely wrinkled, glabrous to densely hairy (especially on the edges) with long white hairs. Leaves bluish-green, sessile or nearly so, undivided, obovate, broadest below the apex, 9–14 × 4–7·5 mm. at maturity, apically rounded in outline, toothed, with teeth 0·5–1 mm. deep, otherwise entire, basally cuneate, varying from glabrous, leathery and dark bluish-green to copiously hairy with long white hairs on both surfaces, the upper surface green, the lower silvery. Flowers usually solitary. Calyx-tube 2 mm. long ; calyx-lobes triangular, 2 mm. long ; epicalyx-lobes lanceolate, shorter than the calyx-lobes. Achenes (1–)2(–3). Fig. 1/1–6, p. 6.

UGANDA. Ruwenzori, Aug. 1938, *Purseglove* 262 ! & Kitandara, Aug. 1953, *Osmaston* 3802 ! & Mt. Gessi glacier-moraine, Mar. 1948, *Mrs. J. Adamson* 19 !
DISTR. **U**2 ; the Belgian Congo ; confined to Ruwenzori.
HAB. Rocky places, moraines and bare patches above upland moor ; 3600–4750 m.

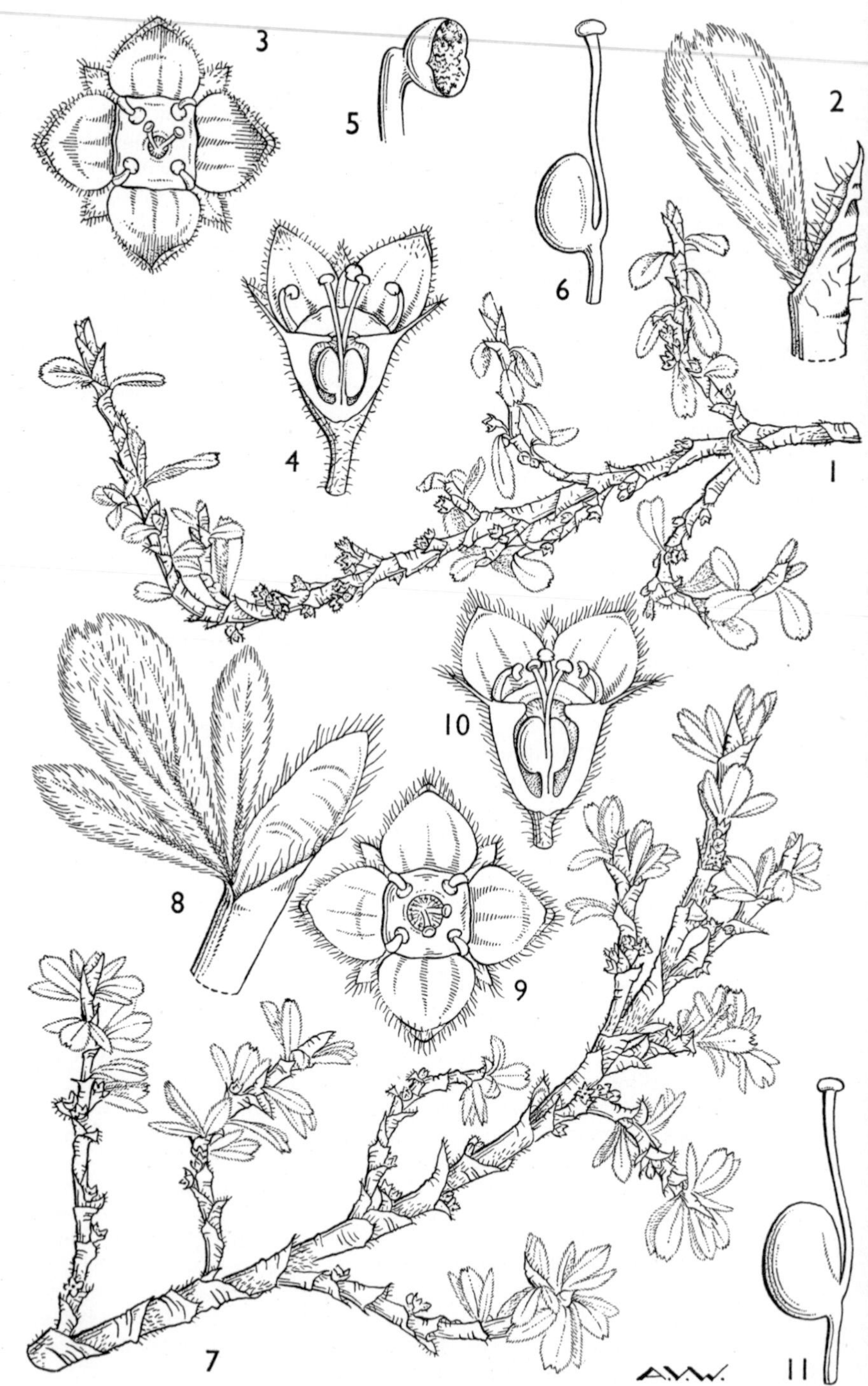

FIG. 1. *ALCHEMILLA SUBNIVALIS*—**1,** part of plant, × 1; **2,** leaf and stipule, × 4; **3,** flower, dorsal view, × 8; **4,** flower in L.S. to show pistil, × 8; **5,** apex of filament with anther, × 32; **6,** pistil, × 16; *ALCHEMILLA ARGYROPHYLLA* subsp. *ARGYROPHYLLOÏDES*—**7,** part of plant, × 1; **8,** leaf and stipule, × 4; **9,** flower, dorsal view, × 8; **10,** flower in L.S. to show pistil, × 8; **11,** pistil, × 16. All from living plants, coll. *Osmaston*, Ruwenzori, Bujuku Valley.

SYN. *A. tridentata* Cort. in Ann. Bot., Roma 6 : 536 (1908). Type : Belgian Congo, Ruwenzori, Valley of the Lakes, *Abruzzi Exp.* (TO, holo., K, photo.!)
A. bequaertii De Wild. in B.J.B.B. 6 : 213 (1921) ; Hauman & Balle in Rev. Zool. Bot. Afr. 24 : 333 (1934) ; F.C.B. 3 : 14 (1952). Type: Belgian Congo, Ruwenzori, Butahu Valley, *Bequaert* 3992 (BR, holo., K, photo.!)
A. microphylla De Wild., in B.J.B.B. 6 : 218 (1921). Type : Belgian Congo, Ruwenzori, *Bequaert* 3760 (BR, holo., K, iso.!)
A. subnivalis Bak.f. forma *nana* Hauman & Balle in Rev. Zool. Bot. Afr. 24 : 336 (1934), *ex descr.* Type : Belgian Congo, Ruwenzori, *Humbert* 8944B (BR, holo.)
A. subnivalis Bak.f. var. *glabrescens* Hauman & Balle in Rev. Zool. Bot. Afr. 24 : 336 (1934), *ex descr.* Types : Belgian Congo, Ruwenzori, *Humbert* 8944A & *Hauman* 440 (BR, syn.)
A. subnivalis Bak.f. var. *glabrescens* Hauman & Balle forma *perpusilla* Hauman & Balle in Rev. Zool. Bot. Afr. 24 : 337 (1934), *ex descr.* Types : Belgian Congo, Ruwenzori, *Hauman* 439 & 430 bis & *Humbert* 8944 (BR, syn.)

NOTE. I follow Hedberg in regarding *A. subnivalis* and *A. bequaertii* as conspecific, such differences as occur being thought to be due to altitude, the latter occupying a lower zone and perhaps not rising above 4500 m. *A. subnivalis* var. *glabrescens* can be separated from the species only in terms of extremes, there being many intermediates.

3. **A. triphylla** *Rothm.* in F.R. 42 : 117 (Apr. 1937) ; Hedb. in Symb. Bot. Ups. 15 (1) : 113 (1957). Type : Belgian Congo, Ruwenzori, *Hauman* 436 (BRLU, holo., BR, K!, P, U, iso.)

A low shrub, 60–120 cm. high. Stems woody, green towards the extremities, sometimes hidden by the stipules, thickly covered with ascending silvery-creamy hairs. Stipules reddish-brown, membranous, usually entire, more rarely apically dentate, wrinkled, glabrous but for a long-ciliate margin, or covered with long ascending hairs. Leaves flabelliform in outline, 3-lobed to the base or nearly so, variable in size, up to 1·5 cm. long and 1·3–2·2 cm. broad, with silvery-creamy hairs covering both surfaces but more silvery beneath ; lobes obovate-cuneate or oblanceolate-cuneate, the lateral as large as the central one or only a little narrower, 3–7-dentate, with teeth up to 2 mm. deep but usually less and the central one the smallest, each with a terminal tuft of hairs ; petioles up to 3 mm. long, but usually shorter, the leaves appearing sessile. Inflorescences 1–2-flowered, short, just emerging from the stipules. Flowers ± 3 mm. long (4 mm. *fide* Rothmaler) ; pedicels up to 1 mm. long. Calyx covered externally with long ascending yellowish hairs ; calyx-lobes ovate-triangular, acute, 1·8–2 mm. long ; epicalyx-lobes lanceolate, shorter than the sepals. Achenes 2.

UGANDA. Ruwenzori, Kigo, Aug. 1931, *Fishlock & Hancock* 57! & Bigo, Jan. 1951, *Osmaston* 3705! & Stuhlmann Pass, July 1951, *Osmaston* 3912!
DISTR. **U2** ; the Belgian Congo ; confined to Ruwenzori.
HAB. Upland moor, preferring damp places, often scrambling over rocks ; 3150–3800 m.

SYN. *A. argyrophylla* Oliv. subsp. *trifolioïdea* Hauman & Balle in Rev. Zool. Bot. Afr. 24 : 332 (1934). Type as *A. triphylla*
A. trifolioïdea (Hauman & Balle) Hauman in Grunne, Hauman, Burgeon & Michot, Le Ruwenzori: 254 (Oct. or Nov. 1937) ; F.P.N.A. 1 : 251 (1948) ; F.C.B. 3 : 13 (1952)

4. **A. argyrophylla** *Oliv.* in Hook., Ic. Pl. 16, t. 1505 (1885) ; Hauman & Balle in Rev. Zool. Bot. Afr. 24 : 353 (1934) ; T.S.K. 58 (1936) ; Hjelmq. in Bot. Not. 109 : 21 (1956) ; Hedb. in Symb. Bot. Ups. 15 (1) : 287 (1957). Type : Tanganyika, Kilimanjaro, *H. H. Johnston* 26 (K, holo.!, BM, iso.!)

A low shrub, 1 m. or more tall, covered except for the stipules with silvery hairs. Branchlets 2–10(–15) cm. long. Stipules reddish-brown, transversely wrinkled, glabrous or with silky marginal hairs. Leaves nearly sessile, deeply 3-lobed ; lobes 4–10 mm. long and 2–5 mm. broad, covered with silvery hairs, the central lobe obovate, obovate-obtriangular to long-obtriangular, or cuneate, broadest at or near the apex, tridentate with teeth up to 4·5 mm. deep ; the lateral lobes entire, linear to linear-elliptic, up to 2 mm. broad, tending to turn inwards slightly at the apex. Inflorescences axillary, 5–7-flowered. Calyx-tube ± 1·5 mm. long ; calyx-lobes triangular, 1·5–1·75 mm. long ; epicalyx-lobes linear-lanceolate, shorter than the calyx-lobes. Achenes 1–2.

subsp. **argyrophylla**

Central lobe of the leaf obovate-obtriangular to long-obtriangular, the terminal teeth exceeding 1·5 and up to 3(–4·5) mm. in depth.

KENYA. Aberdare Mts., Kinangop, Dec. 1930, *Napier* 684 ! ; " Aberdare Range and base of Mt. Kenya ", *Dowson* 105 ! ; Mt. Kenya, Dec. 1943, *Mrs J. Bally in Bally* 3357 !
TANGANYIKA. Kilimanjaro, Jan. 1955, *Verdcourt* 1231 ! & Sept. 1893, *Volkens* 950 ! & Aug. 1930, *B. D. Burtt* 2334 !
DISTR. **K**3, 4 ; **T**2 ; the Sudan (Imatong Mts.).
HAB. Damp places in upland moor and moor-grassland, also on dry, rocky ground, often dominant ; 2250–4500 m.

SYN. *A. argyrophylla* Oliv., *sensu stricto* ; Hedb. in Symb. Bot. Ups. 15 (1) : 111 (1957)
A. keniensis Bak. f. in B.J.B.B. 6 : 217 (1921). Type: Kenya, Mt. Kenya, *Mackinder* (BM, lecto. !, BR, isolecto.)
A. robertii T. C. E. Fries in Arkiv Bot. 18 (11) : 17 (1923). Type : Kenya, Mt. Kenya, *Fries* 1288 (U, holo., BR, K !, iso.)
A. rammellii T. C. E. Fries in Arkiv Bot. 18 (11) : 17 (1923). Type : Aberdare Mts., *Fries* 2401 (U, holo., BR, K !, iso.)
A. argyrophylla Oliv. subsp. *euargyrophylla* var. *typica* Hauman & Balle in Rev. Zool. Bot. Afr. 24 : 354 (1934) ; T.T.C.L. : 473 (1949). Type: as subsp. *argyrophylla*
A. argyrophylla Oliv. subsp. *keniensis* (Bak. f.) Hauman & Balle in Rev. Zool. Bot. Afr. 24 : 355 (1934) ; T.T.C.L. : 473 (1949)
A. argyrophylla Oliv. subsp. *euargyrophylla* Hauman & Balle var. *thorei* Hauman & Balle in Rev. Zool. Bot. Afr. 24 : 355 (1934). Type : Kenya, Mt. Kenya, *Hutchins* 382 (K, holo. !)
A. argyrophylla Oliv. subsp. *euargyrophylla* Hauman & Balle var. *robertii* (T. C. E. Fries) Hauman & Balle in Rev. Zool. Bot. Afr. 24 : 354 (1934)

subsp. **argyrophylloïdes** (*Bak. f.*) *Rothm.* in F.R. 42 : 115 (1937) ; F.C.B. 3 : 13 (1952) ; Hedb. in Symb. Bot. Ups. 15 (1) : 113 (1957). Type : Uganda, Ruwenzori, *Wollaston* (BM, holo. !)

Central lobe of the leaf obovate ; the terminal teeth shallow, up to 1·5 mm. in depth, sometimes absent. Fig. 1/7–11, p. 6.

UGANDA. Ruwenzori, Jan. 1951, *Osmaston* 3707 ! & Aug. 1933, *Eggeling* 1312 ! & Aug. 1938, *Purseglove* 260 !
DISTR. **U**2 ; the Belgian Congo.
HAB. Probably as for subsp. *argyrophylla* ; 2850–3900 m.

SYN. *A. argyrophylloïdes* Bak. f. in J.L.S. 38 : 250 (1908)
A. emarginata De Wild. in B.J.B.B. 6 : 216 (1921). Type: Belgian Congo, Ruwenzori, *Bequaert* 4526 (BR, lecto., K, photo. !)
A. argyrophylla Oliv. subsp. *euargyrophylla* Hauman & Balle var. *argyrophylloïdes* (Bak. f.) Hauman & Balle in Rev. Zool. Bot. Afr. 24 ; 331 (1934)
A. argyrophylla Oliv. subsp. *argyrophylloïdes* (Bak. f.) Rothm. forma *bakeri* Rothm. in F.R. 42 : 115 (1937). Type : uncertain
A. argyrophylla Oliv. subsp. *argyrophylloïdes* (Bak. f.) Rothm. forma *tomentosa* Rothm. in F.R. 42 : 115 (1937). Type : uncertain

5. **A. stuhlmannii** *Engl.* in E.J. 17 : 86 (1893) ; Hauman & Balle in Rev. Zool. Bot. Afr. 24 : 328 (1934) ; Rothm. in F.R. 42 : 118 (1937) ; F.C.B. 3 : 12 (1952) ; Hedb. in Symb. Bot. Ups. 15 (1) : 114 (1957). Type : Ruwenzori, *Stuhlmann* 2409 (B, holo.†, BM, fragment, ? holo. !, K [as *Stuhlmann* 2449], iso. !)

A low shrub, up to 1·5 m. tall, with woody stems and branches covered with long, ascending, whitish hairs. Stipules ovate, obtuse, reddish-brown, membranous, usually glabrous except for a marginal fringe of hairs of variable length. Leaves petiolate, obtriangular in outline, flabellately 3–5-lobed, apically rounded to truncate-rounded, narrowly or broadly cuneate, broader than long, up to 3·5 cm. long and 4·5 cm. broad, silvery with white adpressed hairs beneath, more sparsely hairy and often darker green above ; lobes rarely extending beyond $\frac{2}{3}$ length of the lamina, often less than $\frac{1}{3}$, all of subequal depth on any one leaf, very obtuse or rounded, 3–5(–11)-dentate ; petiole up to 1 cm. long but usually less, thickly hairy. Inflorescence an axillary, rather inconspicuous spiciform raceme, 1–3 cm. long, simple or bifurcated at the base ; bracts ± concolorous with the stipules ; pedicels pilose, ± 1·5 mm. long, not exceeding the bracts. Flowers solitary. Calyx-tube ± 2 mm. long, reddish, thickly hairy with long white hairs ; calyx-lobes broadly ovate-triangular, 2 mm. long, ± acute, hairy as the tube on the outside and glabrous within ; epicalyx-lobes narrowly ovate-lanceolate, 2 mm. long, hairy as the calyx-lobes, and equalling their tips or almost so. Achenes 2(–4).

UGANDA. Ruwenzori, *Geilinger in Bally* 349 *in C.M.* 10981 ! & Butahu Valley, Kitandera, Aug. 1953, *Osmaston* 3781 ! ; Kabamba, July 1951, *Osmaston* 3875 !
DISTR. **U**2 ; the Belgian Congo ; endemic on Ruwenzori.
HAB. Upland moor, often exclusively dominant over rocks and on mossy banks ; 2800–4000 m.

SYN. *A. ruwenzoriensis* Rolfe in J.L.S. 37 : 514 (1906). Type : Ruwenzori, *Scott Elliot* 8109 (K, lecto. !)
A. affinis De Wild. in B.J.B.B. 6 : 211 (1921). Type : Belgian Congo, Ruwenzori, *Bequaert* 3758 (BR, holo., K, photo. !)
A. butaguensis De Wild. in B.J.B.B. 6 : 214 (1921). Type : Belgian Congo, Ruwenzori, *Bequaert* 3903 (BR, holo., K, photo. !)
A. dubia De Wild. in B.J.B.B. 6 : 215 (1921). Type: Belgian Congo, Ruwenzori, *Bequaert* 3758 bis (BR, holo., K, photo. !)
A. stuhlmannii Engl. var. *butaguensis* (De Wild.) Hauman & Balle in Rev. Zool. Bot. Afr. 24 : 329 (1934)

5 × 2. **A. stuhlmannii** *Engl.* × **subnivalis** *Bak. f.*

Leaves simple, or 3-lobed, or with a lobe on one side only, with at least two but usually all three such variants occurring on one plant, up to 15 × 4 mm., dark green above, silvery beneath, with long ± appressed hairs on both surfaces but more thinly so above, the simple leaves oblanceolate and the 3-lobed leaves with lobes subequal or with the lateral lobes much narrower and then the central lobe similar to a simple leaf ; lateral lobe of unilaterally divided leaves usually narrower than the main lobe ; lobes not or scarcely exceeding the middle of the lamina. Flowers not seen.

UGANDA. Ruwenzori, Butahu Valley, Kitandara, Aug. 1953, *Osmaston* 3774B !, 3777 ! & 3813 !
DISTR. **U**2 ; not known elsewhere.
HAB. Upland moor ; 3500–3600 m.

NOTE. This putative hybrid stands in its leaf-characters midway between its two putative parents. Only three gatherings have been seen. The position regarding such irregularly lobed examples is by no means clear.

6. **A. roccatii** *Cort.* in Ann. Bot., Roma 6 : 536 (1908) ; De Wild. in B.J.B.B. 6 : 219 (1921) ; Hauman & Balle in Rev. Zool. Bot. Afr. 24 : 330 (1934) ; F.P.N.A. 1 : 250 (1948) ; F.C.B. 3 : 11 (1952) ; Hedb. in Symb. Bot. Ups. 15 (1) : 117 (1957). Type : Ruwenzori, Valley of the Lakes, *Roccati* (TO, holo.)

A woody, dwarf, much-branched shrub. Main branches short, scarcely exceeding 10 cm. in length with very short lateral branchlets, covered with long ± appressed yellowish hairs, becoming sparser with age and sometimes producing ± herbaceous runners. Stipules brown, membranous or sometimes a little herbaceous at the apex, terminally dentate or rarely entire, ± obscuring the stems of young branches, glabrous but for long hairs on the margins, or with a few similar hairs on the lamina. Leaves petiolate, reniform-rounded, 3–5-lobed, the base truncate to very widely cuneate, up to 1·3 cm. long by 2·3 cm. broad, usually glabrous or nearly so above, covered beneath (rarely above) with silky hairs ; lobes broadly oblong, terminally truncate-rounded, often rather shallow, seldom exceeding half the radius of the lamina, the central one 3–7-dentate, and the lateral lobes usually slightly smaller and 3-toothed ; petiole up to 5 mm. long, covered with hairs. Inflorescence (*fide* Hauman) 2–5-flowered, shorter than the leaves. Calyx-lobes much longer than the epicalyx-lobes. Achenes 1–2.

UGANDA. Ruwenzori, *Humphreys* 539A ! ; Bujuku Valley, Aug. 1933, *Eggeling* 1323A ! ; Kitandara, July 1951, *Osmaston* 3908A !
DISTR. **U2** ; the Belgian Congo ; confined to Ruwenzori.
HAB. Upland moor ; 3600–3750 m.

NOTE. This " species " probably originated as a hybrid of *A. johnstonii* with *A. argyrophylla* subsp. *argyrophylloïdes*. There is, as might therefore be expected, some resemblance with 7, *A. dewildemanii*, but the leaf-lobes are more obovate than in the latter where they are more cuneate, in this respect reflecting in each case the different parent subspecies. *A. roccatii* may however consist of more than one form of the suggested hybrid, the position being uncertain.

7. **A. dewildemanii** *T. C. E. Fries* in Arkiv Bot. 18 (11) : 19 (1923) ; F. R. 42 : 118 (1937) ; Hedb. in Symb. Bot. Ups. 15 (1) : 117 (1957). Type : Kenya, Mt. Kenya, *Fries* 1324 (U, holo. !, BR, K !, iso.)

A low creeping shrub with branches 10–14 cm. long. Stipules reddish, membranous throughout or ± leafy, broadly ovate, apically entire and obtuse to dentate and truncate, the lamina readily wrinkling, with long yellowish hairs on the margin, otherwise glabrous. Leaves rather small, petiolate, flabelliform (? or reniform), 5-lobed, basally truncate (? or very broadly cuneate), 1·1–1·2 × 1·7 cm., the surfaces nearly concolorous, both covered (more thickly beneath) with ± appressed yellowish-silvery hairs ; lobes obovate-cuneate, varying in depth on individual leaves, that is, the sinus between the lateral lobes is shallower (3–5 mm. deep) than that dividing the central lobe from the laterals (7–9 mm. deep), broadest shortly below the apex, terminally 5–7-dentate with the teeth subequal and ± obtuse. Flowers not known.

KENYA. Mt. Kenya, western side, Jan. 1922, *Fries* 1324 !
DISTR. **K4** ; not known elsewhere.
HAB. Upland moor ; 3100 m.

NOTE. This " species ", originally described on rather slender evidence, may be a hybrid between *A. argyrophylla* subsp. *argyrophylla* and *A. johnstonii*, as has been suggested by Hedberg. However the paucity and inadequacy of material make a definite diagnosis extremely difficult.

8. **A. johnstonii** *Oliv.* in Hook., Ic. Pl. 16 : t. 1504 (1886) ; De Wild. in B.J.B.B. 7 : 372 (1921) ; T.T.C.L. : 474 (1949) ; Hjelmq. in Bot. Not. 109 : 21 (1956) ; Hedb. in Symb. Bot. Ups. 15 (1) : 114 (1957). Type : Tanganyika, Kilimanjaro, *H. H. Johnston* 154 (K, holo. !, BM, iso. !)

A low straggling shrub with prostrate stems and erect or procumbent branches, often dominant in carpeting masses ; sometimes (as in *Johnston* 154) with small and densely congested leaves. Stems woody or sometimes ± herbaceous at the extremities, rooting rather infrequently at the nodes, usually reddish-brown, covered (often thickly) with long, ascending or ± spreading, whitish hairs, rarely glabrous. Stipules ± 3–8(–20) mm. long, the lower half (approx.) membranous, the upper half a spreading, foliaceous, lobed limb. Leaves petiolate, often leathery, circular to reniform in outline, 3–5(–7)-lobed, 5–40 mm. long and usually broader than long ; lobes often up to 1 cm. deep, obtuse or rounded, apically entire or dentate, sometimes appearing emarginate owing to the reduced size of the central tooth, often markedly folded along the midribs and with the basal sinus varying from wide to narrow and almost closed. Inflorescences usually short, simple or branched ; pedicels up to 3·5 mm. long, sparsely hirsute with long, ± ascending hairs. Calyx-lobes triangular to ovate-triangular, 1·5–2 mm. long, with a few hairs on the outside and margins ; epicalyx-lobes lanceolate, shorter than the calyx-lobes. Achenes 1(–2).

UGANDA. Toro District : Ruwenzori, Bujuku Valley, Aug. 1933, *Eggeling* 1284 ! ; Kigezi District : Bigo, July 1951, *Osmaston* 3918 ! ; Mbale District : Madangi, Sept. 1932, *A. S. Thomas* 587 !

KENYA. Elgeyo District : Marakwet Hills, July 1935, *Dale* 3401 ! ; Aberdare Mts., Kinangop, Dec. 1930, *Napier* 681 ! ; Mt. Kenya, Girima Valley, June 1933, *G. C. Rogers* 599 !

TANGANYIKA. Moshi District : Kilimanjaro, Aug. 1932, *Greenway* 3153 ! & Feb. 1933, *G. C. Rogers* 569 ! ; Morogoro District : Lukwangule Plateau, Mar. 1953, *Drummond & Hemsley* 1547 !

DISTR. **U**2, 3 ; **K**3, 4 ; **T**2, 6 ; also in the Virunga Mts., and on Ruwenzori in the Belgian Congo.

HAB. Upland moor, moist bamboo-thicket, moor-grassland, on damp ground and often in bogs ; 2400–4260 m.

SYN. *A. cinerea* Engl. in E.J. 19, Beibl. 47 : 31 (1894) ; T.T.C.L. : 474 (1949). Type : Kilimanjaro, *Volkens* 1537 (B, holo. †, BM !, BR, iso.)

A. ulugurensis Engl. in E.J. 26 : 375 (1895) ; T.T.C.L. : 474 (1949). Type : Tanganyika, Uluguru Mts., Lukwangule, *Stuhlmann* 9154 (B, holo. †)

A. geranioïdes Rolfe in J.L.S. 37 : 514 (1906). Type : Ruwenzori, *Dawe* 678 (K, holo. !)

A. ducis-aprutii Cort. in Ann. Bot., Roma 6 : 152 (1907). Type : Ruwenzori, *Giugno* 1906 (TO, holo., K, photo. !)

A. cinerea Engl. var. *uhligii* Engl. in E.J. 46 : 136 (1911) ; F.P.N.A. 1 : 248 (1948) ; F.C.B. 3 : 11 (1952). Type : Tanganyika, Mt. Meru, *Uhlig* 1065 (B, holo. †, BR, iso.)

A. jaegeri Engl. in E.J. 46 : 137 (1911). Type : Tanganyika, Mbulu District, Mt. Loolmalasin, *Jaeger* 849 (B, holo. †)

A. aberdarensis De Wild. in B.J.B.B. 7 : 363 (1921). Type : Kenya, Aberdare Mts., *James* (K, lecto. !)

A. alluaudii De Wild. in B.J.B.B. 7 : 364 (1921). Type : Kenya, Mt. Kenya, *Alluaud* 147 (P, holo., BR, iso., K, photo. !)

A. johnstonii Oliv. var. *crenato-stipulata* De Wild. in B.J.B.B. 7 : 374 (1921). Types : Kenya, Mt. Kenya, *Alluaud* 170 (P, syn.) ; *Orde Browne* (BM, syn. !) ; Hohnel Valley, *Gregory* (BM, syn. !)

A. geranioïdes Rolfe var. *major* De Wild. in B.J.B.B. 7 : 371 (1921), *ex descr.* Type : Belgian Congo, Ruwenzori, *Bequaert* 4514 (BR, holo.)

A. johnstonii Oliv. var. *lindblomiana* Mildbr. in N.B.G.B. 8 : 228 (1922). Type : Elgon, *Lindblom* (S, holo.)

A. lindblomiana (Mildbr.) T. C. E. Fries in Arkiv Bot. 18 (11) : 31 (1923)

A. pseudomildbraedii T. C. E. Fries in Arkiv Bot. 18 (11) : 32 (1923). Type : Kenya, Mt. Kenya, R. Liki, *Fries* 1464 (U, holo. !)

A. pentagona T. C. E. Fries in Arkiv Bot. 18 (11) : 27 (1923). Type : Kenya, Mt. Kenya, *Fries* 1273 (U, holo., K !, S, iso.)

A. geranioïdes Rolfe var. *ducis-aprutii* (Cort.) Hauman & Balle in Rev. Zool. Bot. Afr. 24 : 322 (1934)

A. geranioïdes Rolfe var. *uhligii* (Engl.) Hauman & Balle in Rev. Zool. Bot. Afr. 24 : 324 (1934). Type : as *A. cinerea* var. *uhligii*

A. geranioïdes Rolfe var. *uhligii* (Engl.) Hauman & Balle forma *elongata* Hauman & Balle in Rev. Zool. Bot. Afr. 24 : 325 (1934), *ex descr.* Type : Belgian Congo, Mt. Karisimbi, *Humbert* 8585 (BR, holo.)

A. geranioïdes Rolfe var. *uhligii* (Engl.) Hauman & Balle forma *congesta* Hauman & Balle in Rev. Zool. Bot. Afr. 24 : 325 (1934) ; T.T.C.L. : 474 (1949). Types : Uganda, Mt. Muhavura, *Snowden* 1557 & 1559 (K, syn. !, BR, isosyn.) & Tanganyika, Mbulu District, Mt. Loolmalasin, *B. D. Burtt* 4201 (K, syn. !) & Belgian Congo, Mt. Muhavura, *Scaetta* 203 (BR, syn., K, isosyn. !)

A. geranioïdes Rolfe subsp. *pseudopecten* Hauman & Balle in Rev. Zool. Bot. Afr. 24 : 349, 352 (1934) ; T.T.C.L. : 474 (1949). Type : Tanganyika, Mt. Hanang, *B. D. Burtt* 2280 (K, holo. !)
A. geranioïdes Rolfe subsp. *eugeranioïdes* Hauman & Balle var. *typica* Hauman & Balle in Rev. Zool. Bot. Afr. 24 : 349, 350 (1934). Type : as *A. geranioïdes*
A. geranioïdes Rolfe subsp. *eugeranioïdes* Hauman & Balle forma *major* (De Wild.) Hauman & Balle in Rev. Zool. Bot. Afr. 24 : 322 (1934)
A. geranioïdes Rolfe subsp. *eugeranioïdes* Hauman & Balle var. *ducis-aprutii* (Cort.) Hauman & Balle in Rev. Zool. Bot. Afr. 24 : 349 (1934)
A. geranioïdes Rolfe subsp. *eugeranioïdes* Hauman & Balle var. *uhligii* (Engl.) Hauman & Balle in Rev. Zool. Bot. Afr. 24 : 349 (1934) ; T.T.C.L. : 474 (1949). Type : as *A. cinerea* var. *uhligii*
A. geranioïdes Rolfe subsp. *alluaudii* (De Wild.) Hauman & Balle var. *typica* Hauman & Balle in Rev. Zool. Bot. Afr. 24 : 349 (1934)
A. geranioïdes Rolfe subsp. *alluaudii* (De Wild.) Hauman & Balle var. *pentagona* (T. C. E. Fries) Hauman & Balle in Rev. Zool. Bot. Afr. 24 : 349 (1934)
A. geranioïdes Rolfe subsp. *alluaudii* (De Wild.) Hauman & Balle var. *aberdarensis* (De Wild.) Hauman & Balle in Rev. Zool. Bot. Afr. 24 : 349 (1934). Type : as *A. aberdarensis*
A. cinerea Engl. var. *geranioïdes* (Rolfe) Hauman & Balle ex Robyns in F.P.N.A. 1 : 247 (1948) ; F.C.B. 3 : 10 (1952)
A. cinerea Engl. var. *geranioïdes* (Rolfe) Hauman & Balle forma *major* (De Wild.) Hauman in F.C.B. 3 : 11 (1952)

NOTE. *A. johnstonii* is a very variable species, and I follow Hedberg in synonymizing many forms hitherto regarded as distinct taxa. To these I add also *A. ulugurensis*, which, although offering claim to separation through usually having branched and often more slender and more elongated inflorescences, is nevertheless apparently too closely akin to typical *A. johnstonii* though numerous intermediates in vegetative and floral characters to warrant such treatment. The branched inflorescence is certainly unusual but occurs also in the isotype of *A. cinerea*. *A. ulugurensis* appears to be confined, so far as present records show, to the Morogoro district of Tanganyika, and its recorded altitude range of 2400–2650 m. suggests that it is a lower altitude form. Very similar forms—but almost all with simple inflorescences—occur in Kenya.

9. **A. ellenbeckii** *Engl.* in E.J. 46 : 135 (1911) ; Hauman & Balle in Rev. Zool. Bot. Afr. 24 : 347 (1934) ; Hedb. in Symb. Bot. Ups. 15 (1) : 103 (1957), excl. spec. cit. *Michelmore* 941 !. Type : Ethiopia, Sidamo, *Ellenbeck* 1784 (BR, iso.)

A creeping scrambling or erect herb without central rosettes of leaves, sometimes forming dense masses. Stems often slender, reddish to brownish-green, glabrous to densely pilose with ± ascending, pale brown or whitish hairs. Stipules bilobed, dentate, leafy in the upper part. Leaves petiolate, reniform to rounded-reniform, (3–)5-lobed, 1–1·5(–2·2) cm. long, 1·4–2·5 (–3·6) cm. broad, glabrous to thickly hairy, paler beneath; lobes shallow or extending sometimes to the base (the lateral lobes usually less deeply divided), obovate-cuneate, apically truncate to rounded, dentate (but sometimes extremely shallowly) in the upper part, with the central tooth sometimes vestigial and the basal sinus usually wide, sometimes as much as 180° ; petiole 0·2–1·3(–4) cm. long. Inflorescence simple or much-branched, slender; pedicels glabrous, 0·75–1·5 mm. long; flowers 1–3 together. Calyx glabrous, the tube 1–1·25 mm. long ; calyx-lobes ovate-triangular, rather obtuse, 1–1·2 mm. long ; epicalyx-lobes 0·6–1·2 mm. long, rather variable in width, sometimes equalling the calyx-lobes or nearly so ; disc sometimes with a few hairs, the whole flower otherwise glabrous and sometimes red-tinted. Achenes 1–3.

subsp. **ellenbeckii** ; Hedb. in Symb. Bot. Ups. 15 (1) : 103 (1957), excl. spec. cit. *Michelmore* 941

Leaves with lobes reaching to or commonly beyond the centre of the lamina, sometimes to the base. Calyx-lobes 1–1·2 mm. long ; epicalyx 0·75–1·2 mm. long, thus sometimes equalling the calyx-lobes. Whole plant glabrous or nearly so (rarely with hairy stems and leaves).

UGANDA. Kigezi District: Behungi, Dec. 1933, *A. S. Thomas* 1059 ! & Kanaba Gap, June 1946, *Purseglove* 2071 ! & Sept. 1952, *Lind* 175 !
KENYA. Ravine District : Timboroa, June 1953, *G. R. Williams* 558 ! (hairy form) ; Naivasha District : Mau Forest, Endabarra, Jan. 1946, *Bally* 4825 ! & Bondui, Jan. 1946, *Bally* 4926 !
TANGANYIKA. Masai District : Ololmoti Crater, Sept. 1932, *B. D. Burtt* 4398 !
DISTR. **U**2, 3 ; **K**3, 4 ; **T**2, ?6 ; Belgian Congo (area of Ruwenzori), Ethiopia.
HAB. Upland moor, moist bamboo-thickets, upland grassland, edges of upland rain-forest, often in marshes or bogs and by streams, scrambling over stones, logs, etc. ; 2100–3900 m.

SYN. *A. gracilis* Pax in E.J. 39 : 622 (1907) *nec* Opiz (1838) *nec* Buser (1895), *nom. illegit.* Type : Ethiopia, Shoa, *Rosen* (BRSL, holo.)
A. ellenbeckii Engl. in E.J. 46 : 135 (1911), *sensu stricto*
A. mildbraedii Engl. in Z.A.E. : 226, t. 22 A–D (1911) ; De Wild. in B.J.B.B. 7 : 363 (1921) ; Fries in Arkiv Bot. 18(11) : 23 (1923) ; Hauman & Balle in Rev. Zool. Bot. Afr. 24 : 314 (1934) ; F.P.N.A. 1 : 247 (1948) ; F.C.B. 3 : 8 (1952). Type : Ruanda-Urundi, Rukarara, *Mildbraed* 973
A. brownei De Wild. in B.J.B.B. 7 : 367 (1921). Type : Kenya, Mt. Kenya, *Orde Browne* (BM, holo. !)
A. pickwellii T. C. E. Fries in Arkiv Bot. 18(11) : 24 (1923). Type : Kenya, Mt. Kenya, *Fries* 686 (U, holo., K, iso. !)
A. hillii T. C. E. Fries in Arkiv Bot. 18(11) : 24 (1923). Type : Kenya, Mt. Kenya, *Fries* 1214 (U, holo., K, iso. !)
A. palustris T. C. E. Fries in Arkiv Bot. 18(11) : 25 (1923). Type : Kenya, Aberdare Mts., *Fries* 2312 (U, holo., K, iso. !)
A. granvikii T. C. E. Fries in Bot. Not. 1923 : 54 (1923). Type : Elgon, *Granvik* (S, holo. !)
A. linderi Mildbr. in Journ. Arn. Arb. 9 (11) : 51 (1930). Type : Uganda, Kigezi District, Behungi, *Linder* 2576 (B, holo., GH, K !, iso.)
A. ellenbeckii Engl. var. *hillii* (T. C. E. Fries) Hauman & Balle in Rev. Zool. Bot. Afr. 24 : 347 (1934)
A. ellenbeckii Engl. var. *palustris* (T. C. E. Fries) Hauman & Balle in Rev. Zool. Bot. Afr. 24 : 347 (1934)
A. ellenbeckii Engl. var. *granvikii* (T. C. E. Fries) Hauman & Balle in Rev. Zool. Bot. Afr. 24 : 347 (1934)
A. mildbraedii Engl. var. *villosa* Hauman & Balle in Rev. Zool. Bot. Afr. 24 : 317 (1934), *ex descr.* Type : Belgian Congo, *Lebrun* 3859 (BR, holo.)
A. gracilis Pax var. *hillii* (T. C. E. Fries) Hauman & Balle in B.J.B.B. 14 : 28 (1936)
A. gracilis Pax var. *palustris* (T. C. E. Fries) Hauman & Balle in B.J.B.B. 14 : 28 (1936)
A. gracilis Pax var. *granvikii* (T. C. E. Fries) Hauman & Balle in B.J.B.B. 14 : 28 (1936)
A. ellenbeckii Engl. subsp. *granvikii* (T. C. E. Fries) Hedb. in Symb. Bot. Ups. 15 (1) : 104 (1957)

NOTE. *A. granvikii* has been kept apart by some authors owing to the unusually deep lobing of the leaves. It would seem to be the predominant form on Elgon, but has also been collected in Kenya (Elgeyo District: Marakwet Hills, *Dale* 3402 !; Aberdare Mts., Kinangop, *Napier* 682!) thus there is no very definite regional criterion for distinction. There is a clear range of variation between these very deeply-lobed forms and others such as *Humbert* 7511 ! (Belgian Congo, Kanzibi marsh) where the lobes fail to reach half the radius of the lamina, indeed it is scarcely possible—if *A. ellenbeckii* and *A. granvikii* are kept apart—to know with which *A. hillii* should be synonymized. In the apparent absence of any character correlating with that of the leaf-lobing, it seems advisable to relegate *A. granvikii* to synonymy with *A. ellenbeckii* subsp. *ellenbeckii*.

A. mildbraedii var. *villosa* appears from its type description to be merely a hairy form of typical *A. ellenbeckii*.

subsp. **nyikensis** (*De Wild.*) *R. Grah.* in K.B. 1957 : 406 (1958). Type : Nyasaland, Nyika Plateau, *Henderson* (BM, holo. !)

Leaves with lobes usually not reaching beyond the centre of the lamina and often much shallower. Calyx-lobes 1 mm. long ; epicalyx-lobes 0·6–0·7 mm. long, not reaching the tips of the calyx-lobes. Whole plant, except the flowers, hairy, or the leaves nearly glabrous.

TANGANYIKA. Iringa District : Mufindi West, Jan. 1934, *Michelmore* 941 ! ; Forest Station, Mufindi, Oct. 1947, *Greenway & Brenan* 8286 ! ; Songea District : Matengo Hills, Jan. 1956, *Milne-Redhead & Taylor* 8225 !, 8225A !

DISTR. **T7, 8**; Nyasaland and Northern Rhodesia (Nyika Plateau).
HAB. Upland grassland at edge of upland rain-forest, in bogs and *Phragmites* swamps; 1470–1800 m.

SYN. *A. nyikensis* De Wild. in B.J.B.B. 7: 376 (1921)

10. **A. kiwuensis** *Engl.* in Z.A.E.: 225, t. 21 F–G (1911); Hauman & Balle in Rev. Zool. Bot. Afr. 24: 308 (1934); F.P.N.A. 1: 244, t. 23 (1948); F.C.B. 3: 6, t. 1 (1952); Hedb. in Symb. Bot. Ups. 15 (1): 102 (1957); F.W.T.A., ed. 2, 1: 424 (1958). Type: Ruanda-Urundi, Lake Kalago, *Mildbraed* 1538 (B, holo.†, BR, fragment, ? holo.!)

A low herb usually with basal rosettes with or without creeping, ± slender runners covered with long spreading white hairs, often rooting and developing secondary rosettes at intervals. Stipules commonly 1–2·5 cm. long, united below, free and foliaceous above, the dorsal and ventral sutures being nearly of equal length, thus the aperture scarcely oblique. Rosette-leaves petiolate, rounded to reniform in outline, 5–7(–9)-lobed, 1·8–3(–4·5) cm. long, 2·5–4·5(–7) cm. wide, variably hairy on both surfaces with whitish hairs; lobes usually reaching beyond the centre of the lamina, oblong-elliptic to obovate, terminally rounded, the teeth up to 2 mm. deep and continuing down the sides of the lobes which are entire only towards the base; petiole 3–12 cm. long. Stem-leaves smaller, usually 5-lobed, 1–1·5 cm. long, 1·5–2·5 cm. wide; petiole 1·5–2·5 cm. long. Flowers in 1–4-flowered cymose clusters or in shortly branched panicles; the inflorescence rarely of only 1–2 flowers hidden by the stipules as in *A. cryptantha*; pedicels short, up to 1 mm. long, glabrous or hairy. Calyx-tube ± 2 mm. long, covered with long, spreading white hairs; calyx-lobes 0·75 mm. long, ovate-triangular; epicalyx-lobes exceeding the calyx-lobes, 1–1·25 mm. long, broadly ovate. Achenes 5–8.

UGANDA. Karamoja District: Mt. Debasien, Jan. 1936, *Eggeling* 2698!; Kigezi District: Behungi, Dec. 1933, *A. S. Thomas* 1159!; Ruwenzori, Aug. 1938, *Purseglove* 298!
KENYA. Ravine District: Eldama Ravine, *Whyte*!; Aberdare Mts., Kinangop, Sept. 1951, *Verdcourt* 601!; Mt. Kenya, western side, *Fries* 613A!, 613B!
TANGANYIKA. Moshi District: Lyamungu, Oct. 1943, *Wallace* 1099B!; Ufipa District: Malonje, Apr. 1950, *Bullock* 2780!; Rungwe District: Kiwira R., Lower Fishing Camp, Oct. 1947, *Greenway & Brenan* 8264!
DISTR. **U**1–3; **K**3, 4; **T**2–4, 6–8; Fernando Po, British Cameroons, the Belgian Congo and Southern Rhodesia.
HAB. Open spaces in moist bamboo-thicket and upland rain-forest, in upland grassland, often near streams and on rocky ground; 1250–3000 m.

SYN. *A. adolphi-friedericii* Engl. in Z.A.E.: 225, t. 21 A–E (1911). Type: Ruanda-Urundi, Rukarara, *Mildbraed* 889 (B, holo. †)
A. kiwuensis Engl. var. *kandtiana* Engl. in Z.A.E.: 225 (1911). Type: Ruanda-Urundi, Mt. Niansa, *Kandt* 123 (B, holo. †)

NOTE. This species is distinguished from *A. cryptantha* by the longer, laterally more dentate leaf-lobes, by the tendency to produce persistent primary and secondary rosettes, by the stipules being subequally sutured, and by the stronger, cymose or paniculate inflorescences. There is also an apparent tendency for the urceoles of *A. cryptantha* to be glabrous, whereas in *A. kiwuensis* they are covered with long hairs. Conversely the pedicels of the latter are sometimes glabrous, a feature not noticed in *A. cryptantha*. But none of these characters provides a clear distinction, there being many intermediates, due perhaps either to overlapping of the limits of variation of each or to interspecific hybridity. *Bullock* 2780 approaches subsp. *rhodesica* Hauman & Balle (Rev. Zool. Bot. Afr. 24: 342 (1934)) in having some densely hairy petioles and subglabrous flowers with calyx-lobes nearly equalling the epicalyx, but examples corresponding exactly to the description of this subspecies, which does not seem to be very clearly distinguishable, have not yet been found in our area.

11. **A. microbetula** *T. C. E. Fries* in Bot. Not. 1923: 55 (1923); Hedb. in Symb. Bot. Ups. 15 (1): 107 (1957). Type: Elgon, *Granvik* (S, holo.!)

A small-leaved species, producing tufts of early-withering basal leaves and woody or herbaceous rooting stolons. Stems reddish, covered with long, spreading or ascending, whitish hairs. Stipules entire or with 2–3 teeth, membranous or the teeth (if present) tending to become green and in lush examples ± foliaceous. Leaves on petioles up to 2·5 cm. long, or appearing almost sessile, subcircular in outline, 5-lobed, up to 2·3 cm. diameter, pilose with long silky appressed hairs on both surfaces, often densely but varying to thinly so on the upper surface, generally rather similar to those of typical *A. johnstonii*; lobes 2–4 mm. deep, broadly rounded or ± truncate-rounded and at the apex, jaggedly indented with 5–7 acute, often penicillate teeth ; narrowed (sometimes ± imperceptibly) to the base, typically strongly folded, with the basal sinus usually narrow, sometimes closed or ± so ; petioles adnate for most of their length to the stipules. Inflorescence of 1–2-flowers hidden among the stipules ; pedicels pilose, ± 1 mm. long. Flowers 2·5–3 mm. long. Calyx-tube covered externally with long ascending hairs ; sepals 0·75–1·5 mm. long, ovate-lanceolate or lanceolate ; epicalyx-lobes equal to or slightly longer and narrower than the calyx-lobes, ± lanceolate, the apices penicillate with hairs up to 0·75 mm. long. Achenes 3–12.

UGANDA. Ruwenzori, Aug. 1938, *Purseglove* 261 ! & Nyamgasani Valley, Jan. 1935, *Synge* 154 ! & Namwamba Valley, Jan. 1935, *G. Taylor* 3013 !
KENYA. Elgon, Feb. 1935, *G. Taylor* 3527 ! & Maji ya Moto, May 1948, *Hedberg* 896 !
TANGANYIKA. Moshi District : Kilimanjaro, saddle between Kibo and Mawenzi, June 1948, *Hedberg* 1341 !
DISTR. **U**2, 3 ; **K**3, ?5 ; **T**2
HAB. Upland moor, in sphagnum-bogs, on moraines and in damp gravelly places, often by streams ; 3350–4400 m.

NOTE. *A. microbetula* varies in much the same way as does *A. johnstonii* from **a** dwarf, congested-leaved woody plant to a slender, long-creeping, herbaceous state with stipules ± 3·5 cm. apart, recalling *A. cryptantha*. It is tempting to question whether its origin is in fact *A. johnstonii* × *cryptantha*, but this is scarcely supported by the liberal production of fertile seed, although the number of the carpels is perplexingly variable. It bears certain resemblances to both these species : from the former it is distinguished by having abundantly rooting stolons and tufts of radical leaves (which seem to disappear early), and from the latter by having stolon-stipules whose apices although often dentate and green are sometimes wholly membranous ; also the number of teeth to the leaf-lobes of *A. cryptantha* is greater (9–11 as opposed to 5–7) and the teeth themselves are shallower (up to 1 mm. deep in *A. cryptantha* as opposed to 1·75 mm. in *A. microbetula*) : as regards the last character the fact that the teeth are often deeper in the smaller-leaved *A. microbetula* provides a character whereby the two species may usually be distinguished without difficulty.

12. **A. cryptantha** *A. Rich.*, Tent. Fl. Abyss. 1 : 259 (1847) ; F.T.A. 2 : 377 (1871) ; B.J.B.B. 14 : 8 (1936) ; F.P.N.A. 1 : 246 (1948) ; F.C.B. 3 : 7 (1952) ; Hedb. in Symb. Bot. Ups. 15 (1) : 105 (1957) ; F.W.T.A., ed. 2, 1 : 424 (1958). Type : Ethiopia, Semen, between " Enschedcap " and " Schoata," *Schimper* 566 (P, lecto., BM, K, isolecto. !)

A creeping herb producing basal rosettes of shortly stalked ephemeral leaves and with long, often slender stolons rooting at the nodes and covered with long, whitish or yellowish hairs. Stipules 0·8–1·2 cm. long, united below, free and foliaceous above, the ventral suture not extending as far as that at the rear, thus the aperture is oblique. Leaves reniform, 5-lobed, 1–1·5(–3·5) cm. long, 1·7–3(–4·5) cm. broad, variably densely covered with long hairs on both surfaces, sometimes nearly glabrous ; lobes broadly rounded to broadly obovate, normally shallow, not reaching beyond the centre of the lamina, the sides entire, the terminal teeth subequal (or the central ones smaller) ; petioles slender, hairy as the stems, up to ± 5 cm. long. Flowers axillary, usually 1–2 together, ± concealed by the stipules, either subsessile or on filiform pedicels up to 8 mm. long (elongated, cymose inflorescences as in *A. kiwuensis* rarely developed). Pedicels hairy as the stems. Calyx-tube glabrous or covered with white hairs ; calyx-lobes ovate-triangular, ± 0·75 mm. long ; epicalyx-lobes exceeding the calyx-lobes, broadly ovate, acute, 1 mm. long. Achenes (2–)5–8.

UGANDA. Acholi District : Imatong Mts., Langia, Apr. 1943, *Purseglove* 1408 ! ; Kigezi District : Kachwekano Farm, Jan. 1950, *Purseglove* 3187 ! ; Mbale District : Budadiri, Jan. 1932, *Chandler* 464 !
KENYA. West Suk District : Kapenguria, May 1932, *Napier* 1978 ! ; S. Nyeri/Embu District : Rupingazi R., Mar. 1922, *Fries* 2074 ! ; Kisumu-Londiani District : Tinderet Forest Reserve, July 1949, *Maas Geesteranus* 5393 !
TANGANYIKA. Moshi District : Machame, above the Central Girls' School, Feb. 1955, *Huxley* 95 ! & N. slope of Kilimanjaro, continuation and Loitokitok ridge, Nov. 1932, *G. C. Rogers* 116 ! ; Lushoto District : Mkuzi, about 6·5 km. NE. of Lushoto, Apr. 1953, *Drummond & Hemsley* 2160 !
DISTR. **U**1–3 ; **K**2–5, ?6 ; **T**2–4, 7 ; widely spread on highlands and mountains from Eritrea and the Sudan to the Transvaal ; also in the Cameroons and Madagascar.
HAB. Upland moor, moor-grassland, moist bamboo-thicket and upland grassland, often near streams ; 1300–4050 m.

SYN. *A. tenuicaulis* Hook.f. in J.L.S. 7 : 191 (1864) ; F.T.A. 2 : 377 (1871). Type : Fernando Po, *Mann* 1447 (K, holo. !)
A. holstii Engl. in E.J. 17 : 86 (1893). Type : Tanganyika, W. Usambara Mts., Mlalo, *Holst* 106 (B, holo. † ; U, fragment, ? holo.)

NOTE.* *A. platystigma* Rothm. (in F.R. 42 : 125 (1937)) (syn. *A. mildbraedii* Engl. var. *mauensis* Hauman & Balle in Rev. Zool. Bot. Afr. 24 : 346 (1934)) has been sunk by Balle into synonymy with *A. cryptantha*. It is based on the extremely poor holotype (Kenya, Mau riverside, *Scott Elliot* 6855 (K !)) and although this treatment may be right (so far as can be seen, the epicalyx appears to exceed the calyx-lobes), the specimen is really too inadequate for any degree of certainty as to its characters.

13. **A. gracilipes** (*Engl.*) *Engl.* in E.J. 46 : 129 (1911) ; De Wild. in B.J.B.B. 7 : 342, 348 (1921) ; Hauman & Balle in B.J.B.B. 14 : 32 (1936) ; T. C. E. Fries in Arkiv Bot. 18 (11) : 35 (1923). Type : Kenya, " Abori ", *Fischer* 240 (B, holo.†)

A stoloniferous herb with erect flowering shoots and rosettes of erect, basal leaves. Stems long, rooting at the nodes and producing secondary rosettes at intervals, covered when young with ± appressed, silky-white hairs. Stipules 5–15 mm. long, foliaceous, terminally dentate, the teeth sometimes red-tipped. Rosette-leaves ± reniform, 7–9(–11)-lobed, 2–3·5 cm. long, 3–6 cm. broad, covered on both sides, or mainly on the nerves only, with appressed silky hairs, usually ± silvery below, darker green above ; lobes varying from lanceolate with a subacute apex to obovate-cuneate with a ± rounded apex, not or seldom reaching beyond the middle of the lamina ; petiole up to 21 cm. long, silvery-hairy. Stem-leaves of similar shape, 5–7-lobed, up to 0·5–1·5 cm. long and 1·9–2·7 cm. broad, the lobes variably deeply divided, obovate-rounded to lanceolate-acute, the teeth usually rather jagged ; petiole (2–)4–5(–8) cm. long. Inflorescence simple or branched, rather slender, commonly 10–15 cm. long ; pedicels slender, spreading, often ± filiform, 0·5–1(–5) cm. long, covered with appressed, silky hairs. Calyx-lobes ovate-triangular (1:8–)2–2·5 mm. long, acute ; epicalyx-lobes equalling or slightly exceeding (rarely shorter than) but always narrower than the calyx-lobes, elliptic-lanceolate, (1·8–)2·8–3 mm. long ; calyx and epicalyx covered outside with appressed, silvery hairs, or more rarely glabrous. Achenes 2(–3).

KENYA. Ravine District : Lake Narasha, Oct. 1953, *Drummond & Hemsley* 4805 ! ; Aberdare Mts., Kinangop, Chania Sasumu Dam, Jan. 1953, *Verdcourt* 874 !; Kiambu District : Limuru, June 1909, *Scheffler* 279 !
TANGANYIKA. Masai District : Embagai crater, Feb. 1954, *Eggeling* 6793 !
DISTR. **K**3, 4, 6 ; **T**2 ; Ethiopia, Eritrea.
HAB. Upland grassland, hillsides, river-banks ; particularly in damp places ; 2250–3120 m.

* See also Note under *A. kiwuensis* (p. 14).

SYN. *A. pedata* A. Rich. var. *gracilipes* Engl., Hochgeb. Trop. Afr.: 237 (1892)
A. pedata A. Rich. var. *lovenii* T. C. E. Fries in Arkiv Bot. 18 (11) : 38 (1923). Type: Elgon, *Lindblom* (S, holo. !)
A. lovenii (T. C. E. Fries) T. C. E. Fries in Bot. Not. 1923: 57 (1923)

NOTE. *A. gracilipes* is seemingly closely related to *A. pedata* A. Rich., but three duplicates of the lectotype of the latter (*Schimper* 1166 (K, isolecto. !)) examined have a less silky, less appressed indumentum and more deeply lobed leaves.

14. **A. rothii** *Oliv.*, F.T.A. 2 : 378 (1871) ; De Wild. in B.J.B.B. 7 : 377 (1921) ; Hauman & Balle in B.J.B.B. 14 : 17 (1936). Type : Ethiopia, Shoa, Ankober, *Roth* 157 (K, holo. !)

A rather stout species with a stocky central tuft of leaves, and long creeping stolons. Stems variably shaggy with spreading or ± deflexed, pale brown hairs. Stipules bilobed, membranous, brownish or ± so in the lower part, foliaceous, green and leaf-like in the upper part, terminally dentate, hairy as the leaves. Leaves round in outline, 7–9-lobed, 2–7·5 cm. diameter, the basal sinus usually narrow, often closed or nearly so, covered on both faces (or only very sparsely so above) with white or brownish hairs, sometimes very soft and silky-white ; lobes usually very shallow, up to 2 (–7) mm. deep (the leaves often appearing to be undivided), terminally very broadly rounded to truncate and dentate with 9–13 subacute teeth ; petioles up to 8 cm. long, hairy as the stems. Inflorescence often much exceeding the leaves and up to 22 cm. long ; pedicels shaggy, 3–12 mm. long ; panicles usually open but sometimes the inflorescence dense. Calyx-lobes ovate-triangular, 1·75–2 mm. long ; epicalyx-lobes ovate-lanceolate, shorter and narrower than the calyx-lobes (more rarely of subequal length); calyx and epicalyx shaggy on the outside ; disc (when dry) tending to become black. Achenes 1–2. Fig. 2/1–3, p. 19.

UGANDA. Karamoja District : Mt. Debasien, May 1939, *A. S. Thomas* 2915 !; Mbale District : Elgon, Jan. 1918, *Dummer* 3559 !
KENYA. Elgeyo District : Marakwet Hills, June 1935, *Dale* 3400 ! ; Nakuru/Masai District : Mau, May 1923, *Battiscombe* 1223 ! ; Mt. Kenya, Soames' Camp, Apr 1942, *MacLaughlin in Bally* 2807 !
DISTR. **U**1, 3 ; **K**3, 4, 6 ; Ethiopia.
HAB. Moor-grassland, upper edge of upland rain-forest and moist bamboo-thicket, in rocky and moist situations ; 2700–4000 m.

SYN. *A. steudneri* T. C. E. Fries in Arkiv Bot. 18 (11) : 39 (1923) *fide* Rothm. Type : Ethiopia, Magdala, *Steudner* 922 (B, holo. †)
A. sattimae T. C. E. Fries in Arkiv Bot. 18 (11) : 45 (1923). Type : Kenya. Aberdare Mts., *Fries* 2380 (S, holo. ! K, iso. !)
A. cyclophylla T. C. E. Fries in Arkiv Bot. 18 (11) : 46 (1923) ; Hedb. in Symb. Bot. Ups. 15 (1) : 104 (1957). Type : Mt. Kenya, *Fries* 1322 (U, holo. !)

NOTE. The fragmentary holotype has not been re-dissected, but the carpels appear from other material seen to be distinctly stalked, not subsessile as given in the type description.

There is some variation in the degree of hairiness of the calyx, the less hirsute forms having been described as *A. cyclophylla*—which also usually has 9-lobed leaves. There does not, however, appear to be any clear-cut distinction between our species and the latter, which is here placed in synonymy.

15. **A. fischeri** *Engl.*, Hochgeb. Trop. Afr. : 236 (1892) ; Hauman & Balle in B.J.B.B. 14: 32 (1936). Type: Kenya, "Abori ", *Fischer* (B, holo.†)

A large species with creeping stolons and tufts of large leaves produced at intervals, from which arise erect flowering stems. Stipules pale brown to reddish-brown, membranous or more rarely foliaceous at the tip, glabrous varying to ± densely hairy, sometimes marginally ciliate only, 1–4·5 cm. long, adnate throughout to the petiole. Leaves large, handsome, round to reniform in outline, (7–)9–11-lobed, 3–7 cm. long, 5·5–12 cm. broad, the basal sinus varying from very wide (the leaf-base sometimes nearly truncate) to closed, hairy on both faces but more thickly so and sometimes silvery beneath ; lobes reaching to about the centre of the lamina (more rarely to $\frac{2}{3}$ the radius) but often shallower and ± $\frac{1}{4}$ the radius, oblong to obovate, often rounded above, dentate to the base, terminally sometimes truncate, the teeth jagged ; petioles up to 17 cm. long (? more), hairy as the stem. Flowering stems exceeding the leaves, leafless or with 1–2(–3) foliaceous bracts, covered, sometimes shaggily, with soft, ± spreading or deflexed, white or yellowish hairs ; inflorescence paniculate, usually much-branched, often silvery, with the individual flower-clusters sometimes long, slender and spiciform, usually densely lanate; pedicels very short, less than 1 mm. long. Flowers green (but the sepals tending to redden), 1–2 together in each bract. Calyx-lobes ovate-triangular, 1·5 mm. long ; epicalyx-lobes ovate-lanceolate to lanceolate, equalling or shorter but always narrower than the sepals ; calyx and epicalyx with long, ± appressed hairs outside. Achenes 2–4.

KENYA. Aberdare Mts., Kinangop, Dec. 1930, *Napier* 679 ! & Naivasha–Nyeri track, Oct. 1934, *G. Taylor* 1428 ! ; S. Nyeri District : Mt. Kenya, Sagana Valley, Aug. 1949, *Schelpe* 2766 !
TANGANYIKA. Masai District : Mt. Ela Nairobi, Sept. 1932, *B. D. Burtt* 4172 !
DISTR. **K3**, 4 ; **T2** ; also in mountains of southern Ethiopia.
HAB. Moist bamboo-thicket, upland rain-forest and evergreen bushland ; 2350–3400 m.

SYN. *A. bambuseti* T. C. E. Fries in Arkiv Bot. 18 (11) : 38 (1923). Type : Kenya, Mt. Kenya, *Fries* 699 (U, syn., K, isosyn. !)
A. volkensii Engl. var. *penicillata* De Wild. in B.J.B.B. 7 : 356 (1921) ; Hauman & Balle in B.J.B.B. 14 : 20 (1936). Types : Kenya, Mt. Kenya, *Alluaud* 136 (P, syn., K, photo. !) & Aberdare Mts., Kinangop, *Alluaud* 269 (P, syn., K, photo. !)
A. penicillata (De Wild.) Hauman & Balle in Rev. Zool. Bot. Afr. 24 : 344 (1934)

16. **A. volkensii** *Engl.* in E.J. 19, Beibl. 47 : 30 (1894) ; De Wild. in B.J.B.B. 7 : 354 (1921) ; Hauman & Balle in Rev. Zool. Bot. Afr. 24 : 341 (1934). Type : Tanganyika, Kilimanjaro, near Marangu, *Volkens* 912 (B, holo.†, BM, iso. !)

A herb with central rosette and creeping stolons, at intervals rooting at the nodes and producing tufts of basal leaves and erect flowering stems. Stolons rather slender, sometimes crimson, sparsely to thickly covered with many whitish or pale yellow hairs. Stipules reddish-brown, membranous, 1–1·5 cm. long, glabrous or with scattered hairs, adnate to the petioles throughout. Leaves round in outline, 7–9(–11)-lobed, 2·5–6 cm. diameter, long-stalked, rather sparsely covered with long white hairs, which may be ± confined to the nerves, but varying to thickly and silkily hairy beneath ; lobes reaching to or nearly to the centre of the lamina, obovate with the sides often nearly parallel, terminally truncate or nearly so and often emarginate, the sides and apex finely dentate ; petioles 5–15 cm., brownish-green, hairy as the stems. Inflorescence an open panicle ; the flowers borne 6–8 together in usually congested, capitate fascicles 3–5 mm. in diameter ; pedicels ± 0·5 mm. long. Flowers yellow-green, ± 1·25 mm. long; calyx-lobes ovate-triangular ; epicalyx-lobes oblong-lanceolate, shorter and narrower than the calyx-lobes ; calyx outside and epicalyx covered with long white hairs, or lobes of the calyx and epicalyx marginally ciliate only. Achenes 1–2(–4).

UGANDA. Kigezi District : Mt. Muhavura, Nov. 1948, *Hedberg* 2442 ! ; Mbale District : Elgon, Bulambuli, Nov. 1933, *Tothill* 2464 !, 2341 ! & Sept. 1932, *A. S. Thomas* 531 !
TANGANYIKA. Arusha District : Mt. Meru, Nov. 1901, *Uhlig* 617 ! ; Moshi District: Kilimanjaro, near Marangu, June 1948, *Hedberg* 1138 ! & Bismarck Hill, Feb. 1934, *Greenway* 3827 !

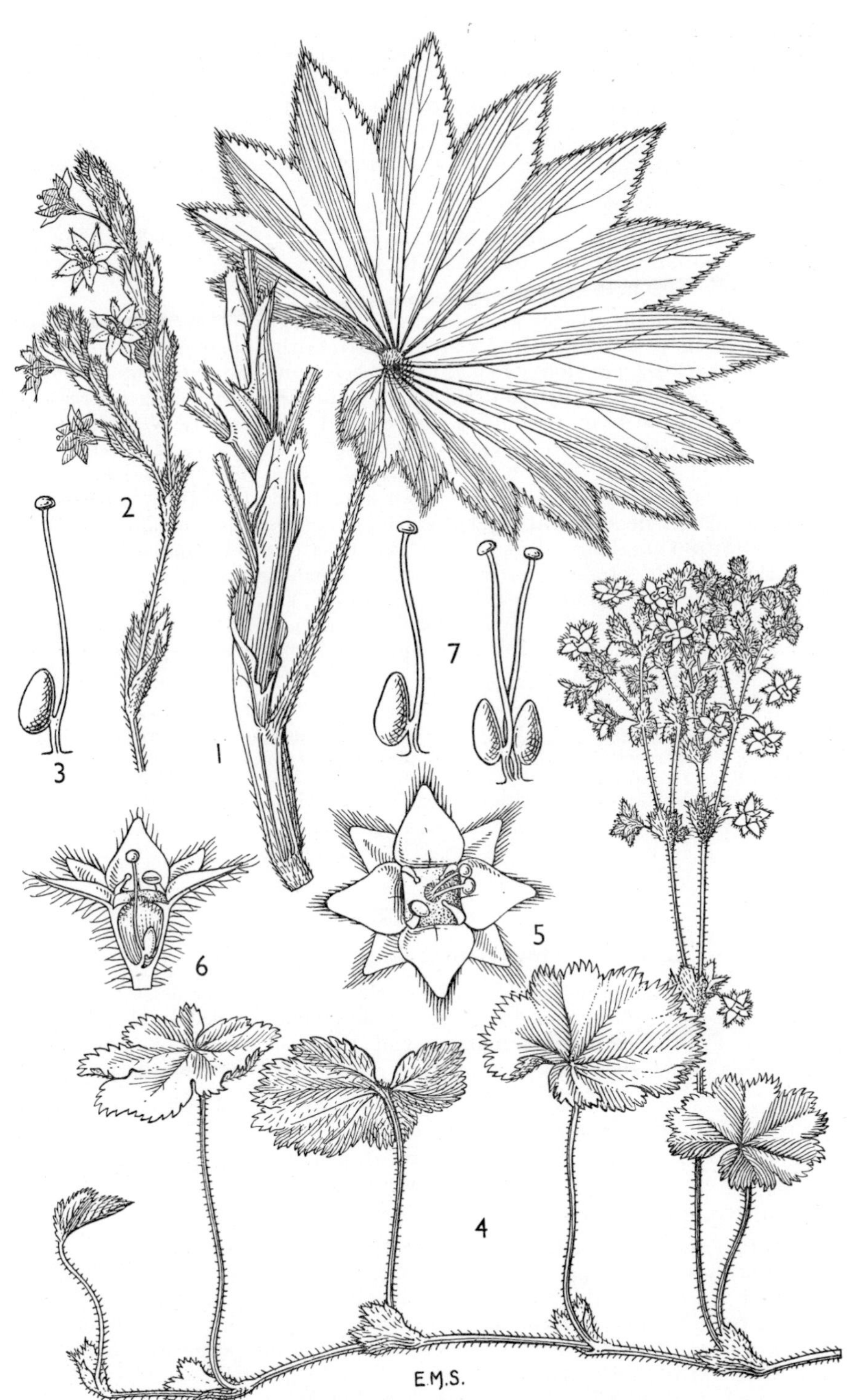

FIG. 2. *ALCHEMILLA HAGENIAE*, from *Napier* 680—**1,** part of stem, with stipules and a leaf, × 1; **2,** part of inflorescence, × $1\frac{1}{2}$; **3,** pistil, × 12; *ALCHEMILLA ROTHII*, from *Battiscombe* 1223—**4,** part of a plant, showing creeping stolon and inflorescence, × 1; **5,** flower, dorsal view, the styles drawn to one side, × 6; **6,** flower in L.S. to show pistil, × 6; **7,** two forms of pistil, × 12.

DISTR. **U**2, 3 ; **T**2; not known elsewhere.
HAB. Upland rain-forest, moist bamboo-thickets, upland evergreen bushland, often covering the ground in damp places ; 1500–2900 m.

SYN. *A. volkensii* Engl. var. *bracteata* De Wild. in B.J.B.B. 7 : 354 (1921). Type : Tanganyika, Kilimanjaro, *Alluaud* 57 (P, holo., K, photo. !)

NOTE. The var. *bracteata* appears to be merely a rather luxuriant form with large leaves and bracts.

17. **A. hagenia** *T. C. E. Fries* in Arkiv Bot. 18 (11) : 33 (1923). Types : Kenya, Aberdare Mts., *Fries* 2277 (U, syn.) & Kinangop, *Fries* 2277A (U, syn., K, isosyn. !)

A large handsome species with decumbent branched stems 1 m. or more long, green or pale pink in colour and covered with appressed, white hairs. Stipules (1–)2–4·3 cm. long, foliaceous or becoming membranous towards the base, greenish-pink in the dried state, glabrous or with long appressed hairs towards the base, terminally bilobed and dentate. Leaves round in outline, 7–17-lobed, up to 10 cm. diameter, glabrous above, covered with appressed, silky-white hairs below (thickly so when young) ; lobes $\frac{1}{4}$–$\frac{1}{3}$ of the radius, triangular (lengthily so when mature) to triangular-lanceolate, often ± acute, dentate throughout with ± appressed, penicillate teeth ; petiole up to 6 cm. long, hairy as the stem, adnate to the stipules for about half their length. Inflorescence paniculate, much-branched, up to 12 cm. long ; pedicels 2–7 mm. long, silkily hairy. Flowers greenish-yellow ; calyx-lobes ovate-lanceolate, acute, 2–3 mm. long ; epicalyx-lobes narrower, but of similar length ; calyx-lobes and epicalyx-lobes covered with appressed hairs, terminally red-tipped and penicillate ; disc reddish. Achenes single. Fig. 2/4–7, p. 19.

KENYA. Aberdare Mts., Kinangop, Dec. 1930, *Napier* 680 ! & Oct. 1934, *G. Taylor* 1282 ! ; N. Aberdare Mts., Melawa Gorge, Mar. 1936, *Meinertzhagen* !
DISTR. **K**3, 4 ; known only from the Aberdare Mts.
HAB. Moist bamboo-thicket and upland evergreen bushland ; 3000–3300 m.

2. FRAGARIA

L., Sp. Pl. : 494 (1753) & Gen. Pl., ed. 5 : 218 (1754)

Perennial stoloniferous herbs. Leaves trifoliolate, densely clustered from a basal rosette. Flowers pentamerous, protogynous. Petals and epicalyx present. Fruit consisting of many, small 1-seeded achenes covering the accrescent, juicy, succulent receptacle.

A temperate and subtropical genus, of which some species and hybrids are widely cultivated and popularly eaten as strawberries.

F. vesca *L.*, Sp. Pl. : 494 (1753) ; Fl. Brit. Ind. 2 : 344 (1878) ; Hegi, Illustr. Fl. Mitt.-Europa, 4: 899, fig. 1157 (1922); Benth. & Hook. f., Brit. Fl., ed. 7 : 140 (1924). Type : uncertain, presumably European

A tufted perennial, emitting from a central rootstock long stolons rooting at the nodes and forming new plants. Stipules red-brown, papery. Leaves petiolate, trifoliolate ; leaflets ovate, oblong, or oblong-ovate, 3–4 × 2–3 cm., obtuse, basally cuneate to ± truncate, sharply serrated with rather few acute and large teeth, green above, with a few appressed hairs or glabrous, whitish below with silky appressed hairs ; petiole densely covered with spreading, silky hairs, commonly 3–6 (but up to 25) cm. long ; petiolules very short, the leaflets often subsessile. Inflorescence an erect, leafless cyme bearing about 5 flowers ; flowering stems up to 15 cm. tall, hairy as the petioles. Flowers 12–18 mm. diameter. Calyx covered externally with appressed silky hairs ; calyx-lobes ovate, acuminate, about 2·75 mm. long ; epicalyx-lobes oblong, acute, nearly equalling the calyx-lobes. Petals white, obovate, ± 4 mm. long. Fruit ovoid or spherical, nodding and red or reddish when ripe.

Kenya. Naivasha District: Aberdare Mts., Naivasha–Nyeri track, Oct. 1934, *G. Taylor* 1404 !
Tanganyika. Morogoro District : unlocalized, *Rounce* 601 ! ; Njombe District : Elton Plateau, 19 Oct. 1956, *Richards* 6589 !
Distr. **K3** ; **T6**, 7 ; a native of Europe, temperate Asia, North America, Madeira, Azores ; status uncertain in our area.
Hab. Upland grasslands and forest edges ; 2400–2850 m.

3. RUBUS

L., Sp. Pl.: 492 (1753) & Gen. Pl., ed. 5: 218 (1754)

Shrubs or scramblers, erect or with long, ± arching sterile shoots (turions) arising from the base and rooting if and when touching the ground. Stems pruinose or not, armed in tropical East African species with prickles, glabrous varying to villous or tomentose, sometimes abundantly covered with ± reddish bristly hairs or stipitate glands. Leaves of tropical East African species petiolate, stipulate, simple, trifoliate, imparipinnate, quinnate or septenate. Flowers perigynous. Calyx-lobes 5, clasping, spreading or reflexing, connate near the base. Petals 5, often much reduced or absent. Stamens many, the filaments commonly glabrous. Carpels many, each containing two ovules, aggregated into a head. Fruit consisting of many one-seeded drupes crowded together on an enlarged or elongated receptacle, which may or may not come away when ripe.

A world-wide genus some of whose sections present great complexity in the number of species, forms etc., due to apogamy, part-sexual reproduction and other causes, the resulting taxa being affected in turn by exposure and other conditions of habitat. Most abundant in temperate regions. The fruits of many species are popularly eaten.

The following account is based entirely on a study of herbarium specimens, with field notes whenever available. Unfortunately the material, although generally adequate numerically, falls markedly short qualitatively owing to the predilection of most collectors to take flowering shoots only—for this reason characters of the turions have had to be abandoned for diagnostic purposes although in a few cases turion leaves, etc. were in fact collected and their features are mentioned. Lack of field-notes similarly necessitates abandonment of the colour of ripe fruit as a character owing to uncertainty over degree of maturity. It must therefore be remembered, when using the following key and descriptions, that leaves and stems refer *only to those on flowering shoots* (unless specifically mentioned otherwise), and that any fruit colour must be accepted with reserve. Inflorescence shape has been described from mature and well-formed examples, but there is considerable variation towards poorer material.

So far as nomenclature is concerned, every effort has been made to establish priorities regarding the various names applied to the *Rubi* of our area, but it is clear that some of our named taxa may more properly be regarded as subservient to earlier names of *Rubi* from other parts of the world. To establish a final nomenclature, even for our area only, requires a prolonged monographic study of the genus on a world basis, and is beyond the scope of this Flora.

Certain authors have described as eglandular those examples where the glands are not obvious and stipitate. In fact most species bear sessile or very shortly stipitate glands on the inflorescence-axes, calyces, and on the leaf-undersurfaces. In the following account sessile or very shortly stipitate glands are not (except sometimes specifically) mentioned in the descriptions, while those described as " stipitate " are at least 0·5 mm. long and are easily visible without a lens when in profusion, and " eglandular " refers to a total lack of glands on the part of the plant thus described.

1. Receptacle markedly elongating with increasing maturity ; carpels very small, produced "in hundreds" ; mature leaves imparipinnate, jaggedly biserrate ; petals white, 1–1·2 cm. long . . . . . 1. *R. rosifolius*
   Receptacle not elongating ; carpels fewer 2
2. Stems thickly covered with eglandular spreading, bristly, dark red-brown hairs 2–3 mm. long ; leaves trifoliolate ; leaflets rounded or very obtuse . . . . 2. *R. ellipticus*
   Stems without eglandular, dark red-brown bristly hairs, such hairs, if present, being gland-tipped (stipitate glands) 3
3. Stipitate glands (including, if present, gland-tipped pricklets) 0·5–1·25 mm. long, covering the stems 4
   Stipitate glands absent or confined to the inflorescence, if present on the stems below the inflorescence, then very few and scattered, not covering the stems 8
4. Calyx-lobes reflexing after anthesis ; leaves simple or trifoliolate, whitish-tomentose beneath ; buds ± globose ; petals not more than 1 cm. long . . . . . 3. *R. steudneri* var. *steudneri*
   Calyx-lobes clasping or spreading after anthesis ; leaves simple, trifoliolate or imparipinnate, green or whitish-tomentose beneath ; buds ± globose to elongate-ovoid ; petals up to 2 cm. long 5
5. Leaves densely whitish-tomentose beneath, trifoliolate ; buds ± globose ; petals 1·2–1·6 cm. long, sometimes broader than long . . . . 4. *R. friesiorum* subsp. *elgonensis*
   Leaves green and hairy beneath, trifoliolate or imparipinnate ; buds ± ovoid-conical 6
6. Leaves mostly imparipinnate ; leaflets acute to long-acuminate, the upper surface not freckled, with or without white hairs and with rather pale stipitate glands (if present) 7
   Leaves all trifoliolate ; leaflets obtuse, the upper surface without white hairs but freckled with many scattered reddish-stipitate glands . . . . . . 7. *R.* sp.
7. Petals up to 9 mm. long ; sepals 5–8 mm. long, whitish-green-tomentose, ± ovate-lanceolate, acuminate . . . . . 5. *R. transvaaliensis* var. *kyimbilensis*
   Petals usually more than 15 mm. long ; sepals 20–25 mm. long, reddish, lanceolate, with a prolonged ± filiform apex . . . . 6. *R. volkensii*
8. Some leaves quinnate 9
   Leaves trifoliolate or imparipinnate (or simple within the inflorescence), not quinnate 10

9. Leaves whitish-tomentose beneath . . . 3. *R. steudneri* var. *aberensis*
   Leaves glabrous beneath or nearly so, the veins often prominent . . . . 3. *R. steudneri* var. *dictyophyllus*
10. Leaves trifoliolate only 11
    At least some leaves imparipinnate 17
11. Calyx-lobes clearly reflexing after anthesis; leaves varying from green and glabrous beneath to whitish-tomentose; buds ± globose; petals longer than the sepals 12
    Calyx-lobes clasping or irregularly spreading after anthesis, but not all clearly reflexed 13
12. Leaves whitish-tomentose beneath . . . 3. *R. steudneri* var. *aberensis*
    Leaves glabrous beneath or nearly so, the veins often prominent . . . . 3. *R. steudneri* var. *dictyophyllus*
13. Petals large, exceeding the sepals. 14
    Petals usually shorter than the sepals or at the most equalling them, sometimes absent 16
14. Leaves ± densely and softly hairy and green beneath, never whitish-tomentose . . 8. *R. keniensis*
    Leaves whitish-tomentose beneath 15
15. Leaves trifoliolate only, whitish-grey-tomentose beneath; lateral leaflets elliptic, usually somewhat obtuse and without a long drawn-out apex . . . . . 4. *R. friesiorum* (aggr.)
    Leaves usually imparipinnate, rarely trifoliolate only, ochreous-grey-tomentose beneath; lateral leaflets ovate-oblong, usually with a long drawn-out very acute apex . 9. *R. runssorensis* var. *runssorensis*
16. Leaves whitish- or fulvous-tomentose beneath; terminal leaflet about 4–9 cm. long; inflorescence cylindrical, exceeding the leaves; petals (if present) 6–9 × ± 4·6 mm.; filaments 2–4 mm. long . . . 10. *R. rigidus*
    Leaves green beneath, hairy on the veins, not tomentose; terminal leaflet not more than 4 cm. long; inflorescence shortly and obscurely cylindrical or corymbose, few-flowered, not exceeding the leaves; petals 3·75 × 2·8 mm.; filaments 1–1·5 mm. long . . . . 11. *R. iringanus*
17. Petals large and showy (1–)1·5–1·8 cm. long; stipules narrowly oblanceolate to broadly obovate-spathulate . . . . . 9. *R. runssorensis*
    Petals not more than 11 mm. long; stipules usually linear, linear-lanceolate to narrowly oblanceolate, often ± filiform 18
18. Mature leaves subglabrous or thinly hairy and green beneath 19
    Mature leaves thickly hairy and grey, greyish- or greenish-white-tomentose beneath 22

19. Petals usually exceeding the calyx ; calyx 7–19 mm. long 20
    Petals usually shorter than the calyx ; calyx 5–8 mm. long . . . . . . 15. *R. pinnatus*
20. Inflorescence many-flowered, ± diffusely cylindrical or broadly pyramidal, exceeding the leafy part of the flowering shoot 21
    Inflorescence few-flowered, short, typically hidden among the leaves, obscurely corymbose, often not more than 5 cm. long ; leaves leathery . . . . . . 14. *R. kirungensis*
21. Leaves with veins markedly prominent beneath, with rather long hairs on the nervature, not tomentose, the lamina very crinkled above, acute or acuminate but sometimes very obtuse ; stems not pruinose, tomentose with long hairs intermixed ; petals rose or pink . . . 13. *R. porotoënsis*
    Leaves with primary nerves prominent beneath, but the intervening veins scarcely or not so, often tomentose on the nervature, the lamina not markedly crinkled above, acute or acuminate ; stems often pruinose, glabrescent below ; petals white, turning magenta . . . 12. *R. scheffleri*
22. Inflorescence many-flowered, exceeding the leafy part of the flowering shoot, if short then not over-topped by leaves 23
    Inflorescence few-flowered, over-topped by leaves ; leaves imparipinnate, 2–3-jugate ; carpels densely tomentose . 18. *R. niveus*
23. Inflorescence short (6–8(–12) cm. long), cylindrical, closely compacted, its axis and branchlets densely hairy with pale fulvous, ± bristly hairs ; leaves usually imparipinnate, 2-jugate ; carpels glabrous . . . . 17. *R. adolfi-friedericii*
    Inflorescence not normally short and if cylindrical then not closely compacted, sometimes broadly pyramidal, its axis and branchlets tomentose or with abundant, pale ± bristly hairs ; leaves imparipinnate or largely trifoliolate ; carpels glabrous or densely hairy 24
24. Petals usually absent ; inflorescence pyramidal, the axis often markedly zigzag, and ± bristly ; leaves usually imparipinnate (2-jugate) ; leaflets ovate to ovate-oblong, more than 1·5 (average 1·7) times as long as broad, acute or acuminate, sometimes with a drawn-out apex . . 16. *R. apetalus*
    Petals usually present, pink ; inflorescence cylindrical, the axis not or slightly zigzag, tomentose, pale ± bristly hairs, if present, not often greatly in evidence ;

leaves usually trifoliolate ; leaflets ovate or ovate-elliptic or obovate, less than 1·5 (average 1·4) times as long as broad, obtuse or ± abruptly acuminate, not usually with a drawn-out apex . . 10. *R. rigidus*

1. **R. rosifolius** *Sm.*, Pl. Ic. Ined. t. 60 (1791) ; Fl. Cap. 2 : 286 (1862) ; Fl. Brit. Ind. 2 : 341 (1878) ; Focke im Bibl. Bot. 72 : 153, fig. 65 (1911) ; C. E. Gust. in Arkiv Bot. 26 (7) : 11 (1934) ; T.T.C.L. : 478 (1949) ; F.W.T.A., ed. 2, 1 : 426 (1958). Type : Mauritius (introduced), *Commerson* (LINN–SM, holo. !)

A handsome scrambling shrub up to 2·5 m. tall, with erect or arching stems covered although scarcely densely so with long spreading white hairs interspersed with scattered or locally numerous amber-green glands. Prickles scattered, small, up to 3 mm. long, straight or abruptly decurved. Leaves imparipinnate (or the uppermost simple), 3–5-jugate (5–7(–11)-jugate on the turions) ; leaflets up to 7·5–9 × 3·5 cm., ovate to oblong-lanceolate, long drawn-out to an acute apex and with a shortly cuneate, rounded or ± truncate base, jaggedly doubly-serrated, the primary serratures saw-like, up to 5 mm. deep, covered above and below with rather long whitish hairs but usually more sparsely so and paler green beneath. Inflorescence consisting of axillary and terminal flowers, borne singly on pedicels 2–4 cm. long. Flowers (2–)2·5–3(–4) cm. in diameter when fully open. Calyx deeply divided ; lobes lanceolate or ovate-lanceolate with a caudate apex, 8–20 mm. long, connate for 3–4 mm. at the base, exceeding the petals. Petals white turning to pink, broadly ovate to subcircular. Carpels not more than 1 mm. long, produced " in hundreds." Receptacle elongating, becoming ellipsoid-cylindrical, up to 2 cm. long (? more) when fully mature. Ripe fruit scarlet, glistening, edible.

UGANDA. Mengo District : Katabi, near Entebbe, Oct. 1950, *Dawkins* 660 ! ; Kampala Plantation, Aug. 1931, *Dep. Agric. Lab. Staff* 2179 !
TANGANYIKA. Moshi District : Weru-Weru Gorge, Feb. 1955, *Huxley* 63 ! ; Lushoto District : W. Usambara Mts., Jaegertal Valley, June 1953, *Drummond & Hemsley* 2966 ! & E. Usambara Mts., Amani, Nov. 1955, *Tanner* 2499 !
DISTR. **U**4 ; **T**2, 3 ; a native of eastern Asia, introduced into Africa, and quite commonly naturalized.
HAB. Edges of upland and lowland rain-forests and plantations, secondary bushland, abandoned cultivations, etc. ; 900–1450 m.

NOTE. The East African examples seen are glandular, but there is variation to an eglandular condition.
When fully ripe, the berries of this species make a delicious dessert fruit.

2. **R. ellipticus** *Sm.* in Rees, Cycl. 30, No. 16 (1819) ; Fl. Brit. Ind. 2 : 336 (1878) ; Focke in Bibl. Bot. 72 : 198 (1911) ; T.T.C.L. : 479 (1949). Type : Nepal, *Buchanan* (LINN–SM, holo. !)

Stems covered, especially densely so above, with spreading reddish-brown, bristly, eglandular hairs 2–4 mm. long. Leaves trifoliolate, the leaflets broadly elliptic and rounded or very obtuse at both ends, up to 5–5·5 × 4 cm., subglabrous and green above, whitish- or yellowish-grey-tomentose beneath, acutely but ± minutely serrated. Inflorescence shortly cylindrical. Calyx 5–6 mm. long ; lobes ± obtusely ovate, 3–4 mm. long. Petals yellow, exceeding the sepals.

TANGANYIKA. Morogoro District : Mzumbi, May 1954, *Semsei* 1712 !
DISTR. **T**6 ; a native of India and Ceylon, introduced here and there in Africa.
HAB. Not recorded.

NOTE. This species is readily distinguished from native species of our area by the dense covering of bristly, reddish-brown *eglandular* hairs on the upper stems.

3. **R. steudneri** *Schweinf.* in Verh. Zool. Bot. Ges. Wien 18 : 669 (1868). Type : Ethiopia, Semen, *Steudner* 921 (B, holo. †)

A scandent shrub up to about 4 m. in height. Flowering shoots ± slender, green to reddish, thickly covered with stellulate-caespitose tomentum varying to glabrous. Prickles small, decurved or ± so. Leaves simple, trifoliolate or quinnate, varying from thinly pilose to glabrous above and from thickly grey-white- or greenish-white-tomentose to glabrous except for the midrib and primary veins beneath ; leaflets obovate, broadly obovate-elliptic or ovate-elliptic, sometimes nearly round, obtuse to ± rounded or sometimes abruptly caudate-acuminate, basally rounded to subcordate, sharply but ± shallowly (1–2·5 mm. deep) serrate or biserrate, the terminal leaflet often larger than the others, 9–15 × 5·6–10 cm. ; venation of non-tomentose forms often prominent and reticulate beneath ; petiolules prickly, glandular or not, those of the terminal leaflets 3–5·6 cm. long, and of the laterals not exceeding 1 cm. long. Inflorescence usually a rather large panicle, pyramidal or ± so, leafless except sometimes at the base, up to about 30 × 17 cm., the axis and branchlets usually grey-green-tomentose, the branchlets spreading at a wide angle or horizontally, aculeolate, with or without stipitate or subsessile glands ; pedicels tomentose and glandular as the branchlets. Calyx 6–9 mm. long, deeply divided into lanceolate oblong-lanceolate or ovate-lanceolate acute sometimes mucronate tomentose lobes basally fused for 1·5–2 mm., strongly reflexed after anthesis. Petals pink, obovate-oblanceolate, exceeding the calyx, 7–10 × 5–8 mm., apically rounded in outline but often notched. Carpels glabrous or with an apical tuft of a few hairs. Fruit orange to dark red.

var. **steudneri**

Flowering stems, inflorescence-axes, peduncles, pedicels and (more sparsely) the calyx (externally) covered with stipitate glands : leaves simple or 3-nate, usually greyish- or greenish-white-tomentose below.

UGANDA. Kigezi District : foot of Kanaba Gap, June 1952, *Lind* 24 ! ; Feb. 1939, *Purseglove* 542 ! ; Mbale District : Elgon, Apr. 1930, *Liebenburg* 1644 !
TANGANYIKA. Morogoro District : Uluguru Mts., above Chenzema, Jan. 1934, *Michelmore* 873 !
DISTR. **U**2, 3 ; **T**6 ; Ethiopia ; probably with a distribution similar to that of var. *aberensis*.
HAB. Moist bamboo-thicket, edges and clearings in upland rain-forest, and in secondary bushland ; 2100–3150 m.

SYN. *R. steudneri* Schweinf. *sensu stricto* ; F.T.A. 2 : 375 (1871) ; C. E. Gust. in Arkiv Bot. 26 (7) : 24 (1934)
R. ulugurensis Engl. var. *goetzeana* Engl. in E.J. 28 : 393 (1900). Type : Tanganyika, Uluguru Mts., Lukwangule, *Goetze* 244 (B, holo.†, LD, photo. !)
R. ulugurensis Engl. var. *adenophloeus* Focke in Bibl. Bot. 72 : 173 (1911). Type : as *R. ulugurensis* var. *goetzeana*
R. adenophloeus (Focke) C. E. Gust. in Arkiv Bot. 26 (7) : 28 (1934) and in K.B. 1938 : 179 (1938) ; T.T.C.L. : 479 (1949)

NOTE. See Note under var. *aberensis* (p. 27).

var. **aberensis** *C. E. Gust.* in Arkiv Bot. 26 (7) : 25 (1934) and in K.B. 1938 : 179 (1938) ; T.T.C.L. : 479 (1949) ; F.C.B. 3 : 22 (1952). Type : Ethiopia, Sidamo, Abera, *Neumann* 41 (B, holo. †)

Stipitate glands absent or extremely few. Leaves simple or 3-nate or quinnate, softly greenish- or greyish-white-tomentose beneath. Fig. 3/3, p. 31.

UGANDA. Karamoja District : Mt. Kadam, Apr. 1959, *Wilson* 737* ; Mbale District : Elgon, Butandiga, Sept. 1932, *A. S. Thomas* 489 ! & Bulambuli, Nov. 1933, *Tothill* 2381 ! & *Saundy & Hancock* 103 !

* This record added by Editor in proof.

Kenya. Nakuru District : Eastern Mau Forest Reserve, Aug. 1949, *Maas Geesteranus* 5951 ! ; Aberdare Mts., Kinangop Forest, Apr. 1943, *Andrews in Bally* 4479 ! ; Kisumu-Londiani District : Tinderet Forest Reserve, July 1949, *Maas Geesteranus* 5527 !

Tanganyika. Morogoro District : Uluguru Mts., Tanana, Jan. 1935, *E. M. Bruce* 626 ! & Morogoro, July 1935, *Rounce* 449 ! ; Iringa District : Mufindi West, Feb. 1932, *St. Clair-Thompson* 628 !

Distr. **U**1, 3 ; **K**3, ?4, 5 ; **T**2, 6, 7 ; Ethiopia and the Belgian Congo.

Hab. Edges of upland rain-forest, upland evergreen bushland and grassland ; 1500–3000 m.

Syn. *R. ulugurensis* Engl. in E.J. 26 : 374 (1899) ; Focke in Bibl. Bot. 72 : 173 (1911) ; C. E. Gust. in Arkiv Bot. 26 (7) : 26 (1934) ; T.T.C.L. : 479 (1949) ; F.C.B. 3 : 22 (1952). Type : Tanganyika, Uluguru Mts., Lukwangule, *Stuhlmann* 9095 (B, holo. †, LD, photo. !)

R. ulugurensis Engl. var. *apricus* C. E. Gust. in K.B. 1938 : 180 (1938) ; T.T.C.L.: 479 (1949). Type : Tanganyika, Iringa District, Mufindi, Kigogo Forest Reserve, *Michelmore* 945 (K, holo. !)

Note. Var. *aberensis* appears to be more widespread than var. *steudneri*, which seems to differ only in the abundance of stipitate glands. In var. *aberensis* these glands are either very few or wholly absent but are replaced by sessile or subsessile glands which are often hidden by the indumentum.

var. **dictyophyllus** (*Oliv.*) *R. Grah.* in K.B. 1957 : 405 (1958). Type : Kilimanjaro, *H. H. Johnston* 183 (K, holo. !)

Stipitate glands apparently absent. Leaves 3–5-nate or the upper ones simple, green beneath, often with prominent, reticulate venation, with hairs on the midrib and veins only, not tomentose.

Uganda. Mbale District : Elgon, Oct. 1916, *Snowden* 444 ! & Oct. 1925, *Snowden* 807 ! & Jan. 1918, *Dummer* 3513 !

Kenya. Aberdare Mts., Kinangop, Dec. 1930, *Napier* 636 ! ; S. Aberdare Mts., Chania Forest, Oct. 1943, *Logie in Bally* 7959 ! ; Kiambu District : Limuru, Uplands, Katamayu Forest, Sept. 1951, *Verdcourt* 609 !

Tanganyika. Moshi District : Lake Chala, Dec. 1932, *Geilinger* 4255 & Kilimanjaro, S. slope between Umbwe and Weru Weru, Sept. 1932, *Greenway* 3216 ! ; Lushoto District : W. Usambara Mts., Shagai Forest, near Sunga, May 1953, *Drummond & Hemsley* 2621 !

Distr. **U3** ; **K**3, 4, ?7 ; **T**2, 3 ; not known elsewhere.

Hab. Edges of and clearings in upland rain-forest and in upland evergreen bushland, lower part of moist bamboo-thickets, hedgerows ; 1800–3000 m.

Syn. *R. dictyophyllus* Oliv. in Trans. Linn. Soc., ser. 2, 2 : 332 (1887) ; Focke in Bibl. Bot. 72 : 172 (1911) ; C. E. Gust. in Arkiv Bot. 26 (7) : 22 (1934) and in K.B. 1938 : 179–180 (1938) ; T.T.C.L. : 478 (1949)

R. keniicola Focke in Ann. Conserv. et Jard. Bot. Genève 20 : 103 (1917) ; C. E. Gust. in Arkiv Bot. 26 (7) : 23, t. 3 (1934). Type : Kenya, Mt. Kenya, *Skene* 115 (G, holo.)

R. rooseveltii Standl. in Smithson. Misc. Coll. 68 (5) : 4 (1917). Type : Kenya, Mt. Kenya, *Mearns* 2371 (US, holo. !)

R. myrianthus Baker var. *ellipticifolius* C. E. Gust. in Arkiv Bot. 26 (7) : 21 (1934), *ex descr.*; T.T.C.L. : 478 (1949). Types : Tanganyika, Usambara Mts., *Holst* 3706 & *Albers* 199 & *Engler* 1311 (B, syn. †)

4. **R. friesiorum** *C. E. Gust.* in Arkiv Bot. 26 (7) : 13, t. 1 (1934). Type : Kenya, Aberdare Mts., Sattima, *Fries* 2603 (U, holo. !)

A suberect shrub with red, glabrous turions and tomentose flowering branches; stipitate glands abundant. Prickles variable in number, red, decurved, up to 3 mm. long, basally dilated and there usually tomentose, otherwise glabrous. Leaves trifoliolate (or simple within the inflorescence), ± coriaceous, dark green above and whitish or whitish-green beneath ; leaflets elliptic to ovate-elliptic, 4·5–8 × 2·5–4·5 cm., the terminal one broadly ovate-elliptic and distinctly larger than the narrower more elliptic laterals, all varying from obtuse to very acute, and from cuneate to rounded basally, but the lateral leaflets often broadest at the centre and narrowing equally to each end, all very shortly puberulous above, thickly and softly whitish-tomentose beneath with prominent midrib and primary veins, and the midrib sometimes aculeolate; margin serrate or biserrate, often ± unevenly so, the serratures up to 2 mm. deep; terminal petiolule up to 2·3 cm. long, those of the lateral leaflets ± 3 mm. long. Inflorescence shortly cylindrical or ± corymbose, the axis sometimes covered with ochreous shaggy hairs, leafless or the rather large flowers borne singly or in pairs in the axils of (often simple) leaves; pedicels 1–2 cm. long, tomentose, prickly, with or without stipitate glands. Calyx tomentose, 9–15 mm. long, deeply divided into lanceolate to ovate or ± triangular lobes 6–12 mm. long with membranous, caudate tips, with or without stipitate glands. Petals pink or white, showy, ± round or broader than long, but abruptly narrowed to the base, exceeding the calyx, 1·2–1·6 cm. long. Carpels varying from densely hairy to glabrous.

subsp. **friesiorum**

Stipitate glands not seen. Leaves elliptic to ovate-elliptic. Ochreous shaggy hairs scarce or absent on the inflorescence-axes. Calyx-lobes lanceolate or oblong-lanceolate, 7–12 mm. long, with a caducous caudate apex. Carpels densely hairy.

KENYA. Aberdare Mts., *Dowson* 108!; Mt. Kenya, Dec. 1943, *Mrs. J. Bally in Bally* 3376! & *Hutchins* 439!

DISTR. **K**?3, 4; not known elsewhere.

HAB. Upper edges of moist bamboo-thickets and of upland rain-forest and evergreen bushland; 3050–3300 m.

SYN. *R. friesiorum* C. E. Gust., *sensu stricto*; C. E. Gust. in K.B. 1938: 177 (1938)
R. friesiorum C. E. Gust. var. *hageniae* C. E. Gust. in Arkiv Bot. 26 (7): 14 (1934) and in K.B. 1938: 178 (1938). Type: Kenya, Mt. Kenya, *Fries* 1274 (U, holo.!)
R. friesiorum C. E. Gust. var. *hageniae* C. E. Gust. forma *albiflorus* C. E. Gust. in Arkiv Bot. 26 (7): 14 (1934). Type: Kenya, Mt. Kenya, *Fries* 1346 (U, holo.!)

NOTE. There is some resemblance between subsp. *friesiorum* and *R. runssorensis* var. *runssorensis* when the leaves of the latter are trifoliolate instead of imparipinnate. The leaflets of the latter are, however, more ovate-oblong, with a longer drawn-out, more acute apex, and the indumentum of the undersurface is more ochreous.

subsp. **elgonensis** (*C. E. Gust.*) *R. Grah.* in K.B. 1957: 406 (1958). Type: Uganda, Elgon, *Dummer* 3541 (K, holo.!)

Stipitate glands abundant on the stems, inflorescence-axes and calyces, or very few or absent. Leaves broadly elliptic to nearly round, and rounded or very broadly cuneate at the base. Inflorescence-axes typically covered with ochreous shaggy hairs. Calyx-lobes broadly ovate, triangular, or oblong-lanceolate, 6–10 mm. long, with a caducous caudate apex. Carpels glabrous or pubescent. Fruit "orange-yellow, very acid."

UGANDA. Mbale District: Elgon, Apr. 1930, *Liebenburg* 1645! & Madangi, Sept. 1932, *A. S. Thomas* 591! & Aug. 1934, *Synge* 913!

DISTR. **U**3; apparently confined to Elgon.

HAB. Clearings in upland rain-forest, moist bamboo-thicket and in moor-grassland; 3300 m.

SYN. *R. friesiorum* C. E. Gust. var. *elgonensis* C. E. Gust. in K.B. 1938: 178 (1938)

NOTE. The flower parts of this handsome bramble are subject to some variation in size; there is also a marked variation in the number of stipitate glands, which are largely absent in the holotype, in *A. S. Thomas* 591, and in *Synge* 913, but very abundant in *Liebenburg* 1645. The affinity is close to subsp. *friesiorum* but the leaves are less narrowed to each end. The stipules of subsp. *elgonensis* are more linear according to available material.

An example from Tanganyika (Mbulu District, north wall of the crater of Oldeani Volcano, *B. D. Burtt* 4226!) was referred by Gustafsson to *R. friesiorum* (K.B. 1938: 177 (1938); T.T.C.L.: 478 (1949)). It is an unusual bramble, eglandular, with trifoliolate leaves and small nearly circular leaflets scarcely more than 3·5 cm. long, in shape more resembling those of subsp. *elgonensis* than subsp. *friesiorum*—with which it however more nearly agrees in its indumentum; it appears in general to be somewhat intermediate between the two. It is not proposed to deal with it as a taxonomic entity until further material becomes available for study.

5. **R. transvaaliensis** *C. E. Gust.* var. **kyimbilensis** *C. E. Gust.* in B.J.B.B. 13 : 27 (1935) and in K.B. 1938 : 185 (1938) ; T.T.C.L. : 481 (1949). Type : Tanganyika, Rungwe District, Mwakaleli, *Stolz* 2286 (K, holo. ! , BM, iso. !)

Habit unrecorded. Axis of flowering shoots stout, tomentose, densely covered with red-stalked glands ; prickles deflexed or ± horizontal, usually straight or slightly hooked, 2–3·5 mm. long. Leaves imparipinnate (2-jugate) or trifoliolate above, bullate ; leaflets broadly ovate, 3·8–6·7 × 2·3–5·3 cm., acute or shortly acuminate, basally ± truncate to cordate, the terminal one largest or ± equalling the lower pair of laterals, green and glabrous or with scattered hairs on the lamina above, rather paler green and more densely hairy beneath with long hairs on the midrib and veins and to a less extent between, unevenly but rather jaggedly serrated with teeth up to 2 mm. deep. Inflorescence obscurely pyramidal, up to 14 cm. long × 7 cm. broad, the axis, branchlets, pedicels and calyx whitish-tomentose, becoming more densely so and more villous upwards and covered in varying density with stipitate glands. Calyx 10 mm. long, deeply divided into ovate-lanceolate, apically caudate lobes ± 7·5 mm. long. Petals entire, very broadly ovate to nearly round, ± 9 mm. long. Carpels densely tomentose.

TANGANYIKA. Rungwe District : Mwakaleli, Nov. 1913, *Stolz* 2286 !
DISTR. **T7** ; known only from a single gathering.
HAB. Moist bamboo-thicket; about 1500 m.

NOTE. As the type of *R. transvaaliensis* is no longer extant (except as a photograph) it is thought advisable to offer var. *kyimbilensis* without distinction from the species.

6. **R. volkensii** *Engl.* in E.J. 19, Beibl. 47 : 30 (1894) ; Focke in Bibl. Bot. 72 : 170 (1911) ; C. E. Gust. in Arkiv Bot. 26 (7) : 18 (1934) ; T.S.K. : 58 (1936) ; T.T.C.L. : 478 (1949). Type : Tanganyika, Kilimanjaro, above Kibosho, *Volkens* 1526 (B, holo.,† BM, K, iso. !)

A handsome, viscid shrub, up to 4 m. tall. Stems brownish-red, densely covered with stipitate and sessile glands intermingled with loose white hairs and scattered prickles. Leaves imparipinnate, or trifoliolate within the inflorescence with the terminal leaflets petiolulate, the lateral ones sessile ; leaflets ovate, ovate-elliptic or ovate-lanceolate, up to 10·5 × 6·5 cm., often long-acuminate, the base rounded to cordate, sharply serrate or biserrate, with very acute teeth, green on both surfaces, pilose and glandular on both surfaces but more so beneath, the young leaves softly villous ; rhachis hairy, glandular and prickly as the stems. Inflorescence loosely paniculate, terminal and axillary, scarcely or not exceeding the leaves. Peduncles and pedicels densely stipitate-glandular. Flowers large, 2·5–3 cm. diameter, apparently somewhat drooping. Calyx densely stipitate-glandular on the outer side, the calyx sometimes with basal pricklets, divided almost to the base into ovate-lanceolate, caudate-acuminate lobes 2–2·5 cm. long. Petals yellow or palely so, sometimes white, broadly ovate, 1·5–2 cm. long, obtuse. Carpels pubescent. Ripe fruit viscid, edible but rather acid, apparently yellow or orange becoming bright red or brownish-red at maturity.

UGANDA. Mbale District : Elgon, Jan. 1918, *Dummer* 3457 ! & Oct. 1923, *Snowden* 819 ! & N. Bugishu, Bulambuli, Sept. 1932, *A. S. Thomas* 558 !
KENYA. Northern Frontier Province : Mt. Nyero, June 1936, *Lady M. Jex-Blake* 33 ! ; Nakuru District : Solai, Richard's Farm, July 1945, *Bally* 4562 ! ; North Nyeri District : Naro Moru, Sept. 1948, *J. G. Williams in Bally* 6417 !
TANGANYIKA. Moshi District : Kilimanjaro, Bismarck Hill, Dec. 1912, *Grote* 3944 ! & near Machame, Jan. 1932, *Moreau* 42 ! & SW. Kilimanjaro, Feb. 1928, *Haarer* 1158 !
DISTR. **U3** ; **K1**, 3, 4 ; **T2** ; Ethiopia.
HAB. Edges and clearings in upland rain-forest and moist bamboo-thicket ; 2100–3450 m.

SYN. *R. mauensis* Engl., in E.J. 46 : 126 (1911) ; Focke in Bibl. Bot. 72 : 170 (1911) ; C. E. Gust. in Arkiv Bot. 26 (7) : 17 (1934). Type : Kenya, Mau escarpment, *S. Baker* 115 (B, holo. †, LD, photo. !)
R. mearnsii Standl. in Smithson. Misc. Coll. 68 (5) : 4 (1917), *non* Elmer (1908), *nom. illegit.* Type : Kenya, Mt. Kenya, *Mearns* 1431 (US, holo. !)

NOTE. *R. volkensii* is widely cultivated for its attractive appearance.
R. mauensis cannot be separated satisfactorily from *R. volkensii* on the characters used by Focke or Engler : for this reason *R. mauensis* is here reduced to synonymy.
It is apparently impossible to establish whether *R. mauensis* was first described by Engler or by Focke—in both cases the year is 1911, though in the first case publication was on 2 May. The benefit of the doubt is here given to Engler.

7. **R. sp.**

Flowering stems thickly covered with red-brown, bristly, glandular hairs interspersed with weak whitish ones. Prickles up to 3 mm. long, mostly slightly hooked. Leaves trifoliolate ; terminal leaflet broadly ovate, ovate-elliptic, or obovate-elliptic, 7–7·5 × 4–6 cm., obtuse, basally truncate to subcordate, 1–1·5 cm. petiolulate ; lateral leaflets smaller and narrower, elliptic or oblong-elliptic, 4·5–6 × 2·5–3·5 cm., mostly obtuse and ± narrowed to each end, subsessile ; all irregularly serrate, the teeth up to 2 mm. deep, ± triangular, their apices sharp due to excurrence of the nerves, green on both sides but paler beneath, rather thickly covered on both sides with stipitate glands like those on the stems. Inflorescence leafy, loosely cylindrical-paniculate, of few, large axillary and terminal flowers ± 2 cm. in diameter when fully open ; the axis and pedicels and calyx-lobes ± densely stipitate-glandular ; pedicels 6–20 mm. long, erect or spreading. Calyx 7–14 mm. long, deeply divided into lanceolate, apically caudate lobes 6–12 mm. long, densely stipitate-glandular outside, green-tomentose within. Petals rose, broadly ovate, 1·1 cm. long, 1 cm. broad, roundly obtuse, basally truncate to a short wedge. Carpels glabrous. Fruit " large, red."

TANGANYIKA. Njombe District : Kipengere Mts., Jan. 1957, *Richards* 7623 !
DISTR. T7 ; not known elsewhere.
HAB. Hillsides in upland evergreen bushland ; 2400 m.

NOTE. This bramble is unlike any other examined during preparation of this account, nor does it seem adequately to fit any consulted description, and affinity with known species is not clear although it would appear to belong to the section *Afromontani* owing to the calyx being cup-like at the base. The available material is all young, and turions are absent ; these factors, added to the fact that there is only one gathering, deter me from describing this bramble yet as a new species.

8. **R. keniensis** *Standl.* in Smithson. Misc. Coll. 68 (5) : 2 (1917) ; C. E. Gust. in Arkiv Bot. 26 (7) : 25 (1934). Type : Kenya, Mt. Kenya, *Mearns* 2325 (US, holo. !, BM, iso. !)

A stout scrambling shrub with reddish-green stems covered, especially thickly above, with loosely-spreading, yellowish hairs interspersed with small, hooked or deflexed prickles scarcely reaching 3 mm. in length. Leaves trifoliolate ; leaflets broadly ovate to ovate-oblong, 8–11·5 × 4–7·5 cm., obtuse or acute, scarcely acuminate, basally rounded to cordate ; terminal leaflet often largest and broader than the lateral ones, all serrate or obscurely biserrate with shallow serratures usually less than 1 mm. deep, dark green and softly hairy above with yellowish, ± appressed hairs, more densely hairy beneath with fulvous or whitish, somewhat shining hairs. Inflorescence a rather stout, pyramidal panicle up to 26 × 18 cm., the branchlets widely ascending, the axis, peduncles and pedicels densely and softly ochreous-villous. Flowers rather large and showy, ± 2·5 cm. diameter. Calyx ochreous- or whitish-tomentose, 9–11 mm. long, deeply divided into ovate or ovate-oblong caudate-acuminate lobes 7–10 mm. long, spreading

Fig. 3. *RUBUS KENIENSIS*, from *Battiscombe* 948—**1,** flowering branch, × $\frac{2}{3}$; **2,** flower, dorsal view, × 1; *RUBUS STEUDNERI* var. *ABERENSIS*, from *A. S. Thomas* 489—**3,** flower, lateral view, × 1; *RUBUS RIGIDUS*, from *Milne-Redhead & Taylor* 8345—**4,** fruit, showing drupes with persistent styles, × 1.

at anthesis. Petals white or pale pink, ± round but abruptly narrowed at the base, 11 × 10 mm. long ; margin wavy, especially at the apex. Carpels densely villous. Fig. 3/1, 2, p. 31.

KENYA. Aberdare Mts., Kinangop, Oct. 1940, *Bally* 1175! & Chania Forest, Oct. 1943, *Logie in Bally* 7960! ; Kiambu District : Limuru, Uplands, Katamayu Forest, Feb. 1939, *Bally in C.M.* 8511!
DISTR. **K**3, 4 ; not known elsewhere.
HAB. Edges of upland rain-forest and on river-banks ; 1950–2670 m.

NOTE. The flowers of this species, like several of the ensuing, are particularly liable to insect attack at a young stage causing the carpels to swell to such an extent that the fruit becomes hard, ± globular, and up to 2·5 cm. in diameter.

9. **R. runssorensis** *Engl.*, P.O.A. C : 190 (1895). Types : Ruwenzori, *Stuhlmann* 2420 & 2453 (B, syn. †)

A scrambling shrub, up to 6 m. tall (*fide* Hauman), producing long hanging branches. Flowering stems tomentose or glabrescent, red or green, pruinose (? always), with many hooked prickles markedly dilated at the base. Leaves imparipinnate (usually 2-jugate) or trifoliolate within the inflorescence ; leaflets broadly ovate, ovate-lanceolate, or elliptic, 4–12 × 2·5–6·5 cm., acute, sometimes acuminate, basally shortly cuneate to subcordate ; the terminal leaflet larger than the lateral ones or about equal to the basal pair, all serrate or biserrate, the teeth rather shallow but acute, commonly 1·5 mm. deep, sparsely pubescent to glabrous above, white- or grey-tomentose varying to green and glabrous beneath (except usually for pilose midrib and nerves) ; terminal petiolule 1–2·5 cm. long ; lateral petiolules not usually exceeding 1 cm. long ; nervature prominent below. Inflorescence consisting of terminal and axillary, often few-flowered, open racemes ; pedicels white-tomentose to glabrous, prickly, 1–4 cm. long. Calyx rather variable in length, 1·2–2 cm. long, deeply divided into lanceolate or narrowly ovate-lanceolate apically acuminate or caudate-acuminate lobes 1–1·8 cm. long, ± whitish-tomentose. Petals showy, white to pale rose, often ± half again as long as the calyx-lobes, sometimes broader than long. Carpels glabrous, or hairy at the apex. Fruit ± 2 cm. long, deep red when young, later black or deep purple, edible.

var. **runssorensis**

Leaves white- or greenish-white-tomentose beneath. Stipules ± narrow, 1·5–4·5 mm. broad.

UGANDA. Ruwenzori, Aug. 1938, *Purseglove* 286! & Aug. 1931, *Fishlock & Hancock* 105! ; Virunga Mts., Mt. Mgahinga crater rim, *Eggeling* 1056!
DISTR. **U**2 ; the Belgian Congo ; confined to Ruwenzori and the Virunga Mts.
HAB. Edges of upland rain-forest and upland moor ; 2400–3750 m.

SYN. *R. runssorensis* Engl., *sensu stricto* ; Focke in Bibl. Bot. 72 : 170 (1911) ; C. E. Gust. in Arkiv Bot. 26 (7) : 15 (1934) ; B.J.B.B. 13 : 267 (1935) ; F.P.N.A. 1 : 236–9 (1948) ; F.C.B. 3 : 21 (1952)
R. kiwuensis Focke in Bibl. Bot. 72 : 171 (1911) ; B.J.B.B. 13 : 268 (1935). Types : Belgian Congo, Mt. Karisimbi, *Mildbraed* 1615, & Mt. Sabinio, *Mildbraed* 1697 (B, syn. †)
R. runssorensis Engl. var. *kiwuensis* (Focke) Engl. in Z.A.E. 224 (1914) ; C. E. Gust. in Arkiv Bot. 26 (7) : 16 (1934) ; F.P.N.A. 1 : 238 (1948) ; F.C.B. 3 : 21 (1952)

NOTE. Gustafsson states that the turions of var. *kiwuensis* are " angled, glabrescent, eglandular, with small prickles slightly compressed from the base, little curved " — an observation which presumably will apply equally to var. *runssorensis*.

It seems to be impossible to establish whether the form of the authority citation for *R. kiwuensis* and *R. runssorensis* var. *kiwuensis* should be as is here given which conforms with F.C.B. in the case of the former.

var. **umbrosus** (*C. E. Gust.*) *Hauman* in F.C.B. 3 : 21 (1952). Type : Belgian Congo, Mt. Mikeno, *Humbert* 8057 bis (P, holo. !)

Leaves green beneath, glabrous, varying to green-tomentose but not white or whitish. Stipules broader (3–8 mm.) than in the preceding variety.

Uganda. Ruwenzori, Aug. 1938, *Purseglove* 237 ! & Bujuku Valley, Aug. 1933, *Eggeling* 1279 ! ; Virunga Mts., Mt. Mgahinga, Aug. 1938, *A. S. Thomas* 2454 !
Distr. **U**2 ; the Belgian Congo ; known only from Ruwenzori and the Virunga Mts.
Hab. Edges of upland rain-forest, upland evergreen bushland ; 2400–3600 m.

Syn. *R. doggettii* C.H. Wright in H.H. Johnst., Uganda Prot. 1 : 325 (1902) ; C. E Gust. in Arkiv Bot. 26 (7) : 65 (1934) and in B.J.B.B. 13 : 268 (1935) and in K.B. 1938 : 178 (1938). Type : Uganda, Ruwenzori, *Doggett* (K, holo. !)
R. runssorensis Engl. var. *kiwuensis* (Engl. ex Focke) Engl. forma *umbrosus* C. E. Gust. in Arkiv Bot. 26 (7) : 16 (1934) ; F.P.N.A. 1 : 239 (1948)

Note. This variety differs from the preceding in the colour of the leaf-undersurface, the white tomentum of var. *runssorensis* not occurring in var. *umbrosus*. Hauman, however, states that shade alone is not, as might be thought, the sole cause of the difference. The stipules of var. *umbrosus* appear to be broader than those of var. *runssorensis*, as indicated, but there is not enough material at the moment for this distinction to be more than a guide.
For comparison with *R. scheffleri*, see under that species (p. 34).

10. **R. rigidus** *Sm.* in Rees, Cycl. 30, No. 5 (1819) ; Fl. Cap. 2 : 287 (1862) ; F.T.A. 2 : 375 (1871) ; Focke in Bibl. Bot. 17 : 174 (1911) ; C. E. Gust. in Arkiv Bot. 26 (7) : 58 (1934) and in B.J.B.B. 13 : 275 (1935) and in K.B. 1938 : 186–7 (1938) ; F.P.N.A. 1 : 242 (1948) ; T.T.C.L. : 480 (1949); F.C.B. 3 : 29 (1952). Type : South Africa, unlocalized (LINN–SM, holo. !)

A very variable, scrambling shrub up to 3 m. tall. Flowering stems greenish-white, usually densely tomentose, sometimes villous or in part so ; prickles variable in number, straight to hooked, 1·75–5 mm. long, basally dilated and tomentose. Leaves commonly trifoliolate, sometimes imparipinnate (2-jugate, the turion leaves apparently frequently so) ; leaflets ovate, ovate-elliptic or obovate but very variable, obtuse or acute or abruptly acuminate, basally rounded to subcordate ; the terminal leaflet usually largest, 3·7–7·5 × 2·7–5·2 cm., or subequal in size to the basal pair in imparipinnate leaves ; all serrate or biserrate, with shallow (sometimes coarse) but usually sharp teeth, rather thinly hairy to glabrous, sometimes very dark green above, whitish-tomentose beneath (varying rarely to green and subglabrous) ; terminal petiolules 0·8–2·5 cm. long, those of lateral leaflets 3–4 mm. long. Inflorescence usually a leafless, rather closely compacted, cyclindrical panicle, the basal branchlets ascending ; axis, branchlets and pedicels whitish-green-tomentose or becoming villous, sometimes fulvous ; pedicels 3–9 mm. long. Flower-buds broadly ovate, cuspidate, 4·5 × 4 mm. Calyx 6–7 mm. long, deeply divided into ovate-lanceolate (sometimes narrowly) or ovate-acuminate, mucronate lobes 5–6 mm. long, greenish-white-tomentose, clasping with maturity. Petals pink to mauve, obovate or obovate-elliptic, 6–9 × 4–6 mm., rounded, apparently caducous. Carpels glabrous, or pubescent particularly at the apex and along the outer side. Fruit edible, ± acid. Fig. 3/4, p. 31.

Uganda. Kigezi District : Ishasha Forest, Kanungu, June 1952, *Lind* 60 ! & Kachwekano Farm, Sept. 1949, *Purseglove* 3098 ! ; Mengo District : Kiagwe, Namanve, Mar. 1932, *Eggeling* 209 !
Kenya. Machakos District : Chyulu Hills, May 1938, *Bally in C.M.* 7900 ! ; N. Kavirondo District : Kakamega, *Carroll* C.6 ! ; Kisumu-Londiani District : Tinderet Forest Reserve, June 1949, *Maas Geesteranus* 5207 !
Tanganyika. Bukoba, *Stuhlmann* ! ; Moshi District : Lyamungu, Mar. 1944, *Wallace* 1194 ! ; Songea District : Matengo Hills, Ngwambo, about 10·5 km. N. of Miyau, Mar. 1956, *Milne-Redhead & Taylor* 8925 !

DISTR. **U**2, 4 ; **K**4, 5 ; **T**1, 2, 4, 7, 8 ; probably throughout most of Africa south of the Belgian Congo and Uganda ; northern limits uncertain.
HAB. Edges of and clearings in upland rain-forest, upland grassland, roadsides, rarely in marshy ground and papyrus-swamp, forming thickets ; 1050–2100 m.

SYN. *R. inedulis* Rolfe in J.L.S. 37 : 514 (1906) ; C. E. Gust. in Arkiv Bot. 26 (7) : 62 (1934) and in B.J.B.B. 13 : 275 (1935) and in K.B. 1938 : 186 (1938) ; F.P.N.A. 1 : 242 (1948) ; T.T.C.L. : 480 (1949) ; F.C.B. 3 : 31 (1952). Types : Uganda, Buddu, *Brown* 133 (K, syn. !) & Koki, *Dawe* 388 (K, syn. !)
R. atrocaeruleus C. E. Gust. in Arkiv Bot. 26 (7) : 54 (1934) and in K.B. 1938 : 185 (1938). Type : Kenya, Mt. Kenya, *Fries* 1802A (U, holo. !)
R. rigidus Sm. var. *discolor* Hauman in B.J.B.B. 22 : 93 (1952) ; F.C.B. 3 : 30 (1952), as " *concolor* ". Type : Uganda, Kigezi District, Kisoro, *Gesquière* 5682 bis (BR, syn. !)

NOTE. Certain specimens (Uganda, Mengo District : Mukono Hill, *Dummer* 2693 (K !) ; Kabulasoke, Gomba, *Maitland in Liebenburg* 1278 (K !)) have markedly thick pedicels and more diffusely spreading panicles than is usual for *R. rigidus*. Possibly they, with other examples from scattered localities in Nigeria, Portuguese East Africa and Belgian Congo, are hybrids. Unfortunately, however, these examples are diseased, as shown by leaf discoloration, and it is thought advisable for this reason to omit them from further consideration in the present account.

11. **R. iringanus** *C. E. Gust.* in K.B. 1938 : 184 (1938) ; T.T.C.L. : 479 (1949). Type : Tanganyika, Iringa District, Kigogo Forest Reserve, *Michelmore* 948 (K, holo. !)

Apparently a dwarf bramble with the flowering stems covered, especially thickly above, with ochreous, ± spreading hairs interspersed with many small, hooked or straightly deflexed prickles up to 2 mm. long. Leaves trifoliolate (or simple within the inflorescence) ; leaflets broadly ovate or obovate, 2·8–4 × 2·4–3·6 cm., or the lateral ones more elliptic or obovate-elliptic and not exceeding 3 × 1·8 cm., acute to very obtuse or rounded, basally cuneate to rounded, with serratures up to 2 mm. deep, all jagged and sometimes markedly so, green on both surfaces but darker above with ochreous hairs and more densely hairy on the midrib and nerves below, with the intervening tissues glabrous. Stipules linear or linear-lanceolate. Inflorescence apparently leafy, short and congested, 6 × 3 cm., of obscure shape, consisting of a few, small, axillary flowers borne singly on pedicels up to 5 (?–10) mm. long. Calyx 7–8 mm. long, deeply divided into ovate-lanceolate, caudate-acuminate lobes ± 6·5 mm. long, covered with long ochreous hairs on the outer side, tomentose within, slightly aculeolate, apparently clasping. Petals white, 3·75 × 2·8 mm., broadly ovate, somewhat truncate. Carpels glabrous. Fruit about 1·2 cm. long, " red when unripe and perhaps also when ripe."

TANGANYIKA. Iringa District : Mufindi, Kigogo Forest Reserve, Jan. 1934, *Michelmore* 948 !
DISTR. **T**7 ; known only from a single gathering.
HAB. Upland rain-forest ; probably about 1800 m.

12. **R. scheffleri** *Engl.* in E.J. 46 : 125 (1911) ; C. E. Gust. in Arkiv Bot. 26 (7) : 43 (1934) and in K.B. 1938 : 183 (1938) ; T.T.C.L. : 481 (1949). Type : Kenya, Kiambu District, Limuru, *Scheffler* 329 (B, holo. †, BM, K, iso. !)

A scrambling shrub, 3–5 m. tall. Flowering branches reddish, fulvous-pilose and sometimes thickly so above, glabrescent and pruinose (? always) below ; prickles numerous, ± stout, hooked, glabrous, up to 5 mm. long. Leaves imparipinnate (2–3-jugate) or the upper ones trifoliolate ; terminal leaflets broadly ovate, up to 6·5 × 3·6 cm., the lateral ones rather smaller and narrower, ovate to ovate-lanceolate, 4·5–6 × 2·3–3·2 cm. ; all acute or acuminate, basally rounded, sharply and irregularly serrate, the nerves often markedly excurrent, green on both surfaces but paler below, glabrous or with pubescent midrib and nerves on either or both surfaces and sometimes with scattered, ± appressed hairs on the lamina above ; lateral nerves straight or arcuate. Inflorescence usually a rather restricted, largely or wholly leafless, cylindrical panicle, but sometimes more diffuse, commonly up to 17 × 5 cm., the peduncles and pedicels usually spreading at an angle of 90° or a little deflexed ; flowers solitary or in pairs ; the branchlets, peduncles (if present) and pedicels grey-green-tomentose and usually very aculeolate. Calyx tomentose, often purplish on the outside, becoming marginally greener, 9–10(–15) mm. long, deeply divided into oblong-lanceolate or lanceolate caudate-acuminate lobes 7–8 mm. long, which spread or reflex after maturity. Petals white, " turning magenta," obovate or obovate-elliptic, 6·5–8 × (4–) 6–6·5 mm., notched or entire. Fruit tomentose, often densely so, orange-red or black (? when mature), edible.

UGANDA. Ruwenzori, Aug. 1938, *Purseglove* 224 !
KENYA. North Nyeri District : Naro Moru, Sept. 1948, *J. G. Williams in Bally* 6418 !; Aberdare Mts., Chania Forest, Oct. 1943, *Logie in Bally* 7957 ! ; Meru, July 1944, *Mrs. J. Bally in Bally* 3998 !
TANGANYIKA. Masai District : Mt. Ololmoti, Sept. 1932, *B. D. Burtt* 4374 ! ; Arusha District : Mt. Meru crater, Oct. 1932, *B. D. Burtt* 4152 ! ; Morogoro District : Uluguru Mts., Lukwangule, Jan. 1934, *Michelmore* 897 !
DISTR. **U**2 ; **K**?3, 4 ; **T**2, 3, 6 ; not known elsewhere.
HAB. Upland rain-forest margins and glades, moist bamboo-thicket, upland evergreen bushland ; 1650–3150 m.

SYN. *R. kirungensis* Engl. var. *pilosior* C. E. Gust. in Arkiv Bot. 26 (7) : 39 (1934). Type : Kenya, Mt. Kenya, *Fries* 582 (U, iso. !)
R. stuhlmannii Engl. var. *aberdarensis* C. E. Gust. in Arkiv Bot. 26 (7) : 43 (1934) and in K.B. 1938 : 183 (1938). Type: Kenya, Aberdare Mts., *Fries* 2295 (U, iso. !)
R. kingaënsis Engl. var. *pubescens* C. E. Gust. in Arkiv Bot. 26 (7) : 48 (1934) and in K.B. 1938 : 184 (1938) ; T.T.C.L. : 480 (1949). Type : Tanganyika, Uluguru Mts., Mkambuku, *von Brehmer* 958 (B, holo. †)

NOTE. *R. kirungensis* var. *pilosior* seems to be a form of *R. scheffleri* with terminal leaflets rather broader than usual and resembling more those of *R. rigidus*.

It is possible, from a photograph of its holotype (LD !), that *R. stuhlmannii* Engl. (E.J. 26 : 374 (1898) ; Focke in Bibl. Bot. 72 : 177 (1911) ; T.T.C.L. : 481 (1949)) is conspecific with *R. scheffleri* but in the absence of any type specimens it is not possible to be sure. For this reason the two cannot be safely synonymized although *R. stuhlmannii* var. *aberdarensis* can be confidently reduced to synonymy with *R. scheffleri*. It should, however, be remembered that *R. stuhlmannii* antecedes *R. scheffleri* and must take priority if the two are ever synonymized at specific level.

R. scheffleri seems to lie midway between *R. pinnatus sensu lato* and *R. runssorensis* var. *umbrosus*. From the former it differs chiefly in its larger flowers, particularly the petals and longer calyx-lobes. But the isotype in the Kew herbarium has petals only 6·5 mm. long, as against 8–9 mm. as claimed in the type description. *R. scheffleri* is not always easy to tell clearly from certain forms of *R. kirungensis*, of which var. *glabrescens* seems to stand midway (as do the Tanganyika records given above) between both in certain characters, although typical *R. kirungensis* is distinguished apart easily enough (see additional observations under the latter sp.). It is scarcely possible to systemize more exactly these apparently closely-affiliated taxa without recourse to further field work.

13. **R. porotoënsis** *R. Grah.* in K.B. 1958 : 107 (1958). Type : Tanganyika, Mbeya District, Poroto Mts., *Richards* 9700 (K, holo. !)

A scrambling shrub. Flowering stems obscurely or obtusely angled, tomentose, with or without intermixed long spreading hairs (sometimes in preponderance), bearing ± abundant straight or decurved prickles 1·5–4 mm. long. Leaves imparipinnate or the upper ones trifoliolate or simple, ovate ovate-oblong, or oblong-elliptic, 4–8 × 2·2–5·5 cm., varying from very acutely acuminate to obtuse, basally broadly cuneate to subcordate, sharply and irregularly serrate, the lamina markedly crinkled due to depression of the nerves and veins, rather sparsely pilose above with long hairs, and with similar hairs on the nervature beneath, varying to more densely hairy on both surfaces. Inflorescence a nearly leafless pyramidal panicle, up to 30 × 20 cm., with the axes and peduncles hairy as the stems; pedicels and calyx-lobes greenish-white-tomentose. Calyx-lobes lanceolate or ovate-lanceolate, 5–19 mm. long, variably long-caudate. Petals rose or pale pink, 9–11 mm. long, 8–10 mm. broad, terminally ± rounded. Carpels tomentose.

TANGANYIKA. Mbeya District: Poroto Mts., near Mbeya, Feb. 1934, *Michelmore* 965!; Rungwe District: Ngozi Forest, Oct. 1956, *Richards* 6485! & N. Ngozi Forest, Oct. 1932, *Geilinger* 2886!
DISTR. **T7**; not known elsewhere.
HAB. Edges of upland rain-forest; 2100–2340 m.

SYN. *R. apetalus* Poir. var. *roseus* C. E. Gust. in K.B. 1938: 182 (1938); T.T.C.L.: 480 (1949). Type: Tanganyika, Poroto Mts. near Mbeya, *Michelmore* 965 (K, holo.!)

14. **R. kirungensis** *Engl.* in Goetzen, Durch Afr.: 378 (1895) and in Z.A.E.: 224, t. 20 (1911); C. E. Gust. in Arkiv Bot. 26 (7): 39 (1934) and in B.J.B.B. 13: 270 (1935) and in K.B. 1938: 184 (1938); F.P.N.A. 1: 240 (1948); F.C.B. 3: 23 (1952). Types: Belgian Congo, Virunga Mts., *Goetzen* 69 (B, syn. †) & 73 (B, syn. †, LD, photo.!)

A mostly glabrous straggling typically very prickly shrub up to 2 m. tall. Flowering stems red or greenish-red, glabrous or pubescent above when young; prickles small, 1–3 mm. long, straightly deflexed or slightly hooked, glabrous. Leaves imparipinnate, or trifoliolate, or simple; leaflets typically small and crinkled, ± leathery, and with pronounced venation beneath; the terminal leaflet often the largest, ovate, ovate-elliptic or obovate, 3(–7·5) × 2·5(–5) cm.; the lateral leaflets usually ovate-elliptic; all acute varying to obtuse or abruptly acuminate, basally rounded to subcordate, serratures typically rather few, obtuse or acute with excurrent veins, green but paler beneath, glabrous or with a few, long, ± appressed hairs on the lamina above and a pubescent midrib beneath; terminal petiolule up to 1·5 cm. long; lateral petiolules very short or up to 4 mm. long, aculeolate, glabrous. Inflorescence very shortly cylindrical, few-flowered, leafy to the apex, the flowers ± hidden amongst leaves; pedicels usually very prickly, glabrous, varying to ± floccose, rarely with a few stipitate glands. Flowers 1·5(–2) cm. across. Calyx 7–10 mm. long, tomentose throughout or becoming glabrous and fleshy on the outside except for the margins of the lobes, sometimes with stipitate glands; calyx-lobes lanceolate, acute or caudate-acuminate, 6–8 mm. long. Petals rose-pink or white, ± rounded, 7–10 × 6·5–7 mm. Fruits glabrous or apically pubescent. Ripe berries apparently black, edible, ? rather acid.

UGANDA. Kigezi District: Mts. Muhavura–Mgahinga saddle, Sept. 1946, *Purseglove* 2143! & Mt. Mgahinga, Aug. 1938, *A. S. Thomas* 2449! & Mt. Muhavura, Oct. 1947, *J. G. Williams in Bally* 5052!
TANGANYIKA. Rungwe District: Upper Kiwira River, May 1938, *MacInnes* 408!
DISTR. **U2**; **T7**; the Belgian Congo (Virunga Mts.).
HAB. Moist bamboo-thicket, exposed lava-outcrops, upland grassland and evergreen bushland; 1950–3360 m.

SYN. *R. goetzenii* Engl. var. *glabrescens* Engl. in Mildbr., Z.A.E.: 224 (1914). Type: Belgian Congo, Mt. Ninagonga [Nyiragonga], *Mildbraed* 1385 (B, holo.)
R. kirungensis Engl. var. *glabrescens* (Engl.) C. E. Gust. in Arkiv Bot. 26 (7): 40 (1934) and in B.J.B.B. 13: 270 (1935); F.P.N.A. 1: 240 (1948)

NOTE. Although readily recognisable in its small-leaved typical form, the larger, shade-grown forms (var. *glabrescens*) appear in some cases to suggest affinity with *R. scheffleri* or *R. runssorensis* var. *umbrosus*. It is these forms (e.g. Uganda, Mt. Muhavura, *Eggeling* 1051! and pass between Mt. Mgahinga and Mt. Sabinio, *G. Taylor* 1903!)

which show a tendency for the calyx-lobes to become partially glabrous on the outside, and to bear stipitate glands, and which have rather longer filaments and styles. Whether or not these forms should be kept taxonomically distinct from typical *R. kirungensis* is difficult to say, but in view of intermediates which clearly connect the one form with the other they are here offered in synonymy with the species.

In addition to these forms, certain apparently different Ruwenzori gatherings have been examined (Bujuku Valley, 1931, *Fishlock & Hancock* 161 ! ; Nyamgasani Valley, 1935, *Synge* 1479) which although having a somewhat distinct appearance seem also to be connected with *R. kirungensis* through intermediates and, for want of further material and further field observation, are included under this species. They may be recognized by their larger lanceolate leaves, up to 11 cm. long, and to about 6 cm. wide with long drawn out very acute apices, rounded bases, and rather few deeply sculptured arcuate nerves, and in some examples by an abundantly stipitate-glandular inflorescence. It is possible that these brambles may constitute a separate species or alternatively may represent a form of *R. kirungensis* influenced by different conditions of habitat. They have some resemblance to *R. runssorensis* var. *umbrosus*, but their petals are smaller, and the nerves of the leaves are fewer and more arcuate. The two examples quoted were found at 2550 and 3300 m. altitude respectively. *G. Taylor* 2938 (Ruwenzori, Namwamba Valley) seems to be intermediate between the arcuately-nerved material and normal *R. kirungensis*.

The specimen cited from Tanganyika is a rather perplexing example which generally resembles *R. kirungensis* but which has a more extended terminal leafless inflorescence. Being a unicate it is difficult to know whether it is abnormal, or whether some other species is involved. Despite its isolated occurrence it is referred here to *R. kirungensis* temporarily.

15. **R. pinnatus** *Willd.*, Sp. Pl. 2 : 1081 (1799) ; F.T.A. 2 : 374 (1871). Type : St. Helena, collector unknown (B, holo.)

A variable sometimes scrambling widely spreading shrub up to 6 m. tall. Flowering shoots with rather slender, greenish-red or red, subglabrous or tomentose stalks ; prickles straight or hooked, glabrous or nearly so, up to 6 mm. long but often considerably less, sometimes congestedly numerous. Leaves imparipinnate (2–4-jugate) or the uppermost trifoliolate ; leaflets rather small, the terminal one often largest, ovate to ovate-lanceolate, 3–6·5 × 1·5–3·3 cm., acute or acuminate, basally rounded to cordate, sharply and sometimes ± deeply (up to 3 mm.) serrate or biserrate, green on both surfaces but paler beneath, usually glabrous but varying to thinly hairy, and usually with pubescent midrib and nerves on the under side. Inflorescence a rather lax panicle varying from ± slender and cylindrical to stout and broadly pyramidal, up to 35 × 20 cm., with the lowest branchlets 12 cm. long or more ; pedicels slender, 7–13 mm. long, tomentose, commonly aculeolate. Calyx 5–8 mm. long, tomentose, rarely with a few stipitate glands or gland-tipped pricklets, deeply divided into greenish lobes 4–6·5 mm. long, irregularly spreading after anthesis. Petals (if present) pink or white, ovate, obtuse, 4 × ± 2·5 mm. Carpels densely tomentose varying to glabrous. Fruit " yellow, red or rosy," but probably black when ripe, edible.

var. **pinnatus**

Prickles of flowering shoots not more than 3 mm. long. Inflorescence eglandular, ± narrowly cylindrical, not pyramidal, up to 14 cm. long but scarcely more than 4 cm. broad, the basal branchlets up to 4·5 cm. long. Carpels glabrous or nearly so.

Kenya. Kiambu District : Limuru, Feb. 1915, *Dummer* 1615 ! ; Kericho District : SW. Mau Forest Reserve, Aug. 1949, *Maas Geesteranus* 5688 !

Tanganyika. Lushoto District : W. Usambara Mts., Nguelo [? Nkalai], *Scheffler* 70 ! & Shagai Forest, May 1953, *Drummond & Hemsley* 2569 ! ; Morogoro District : Uluguru Mts., Kitundu, Oct. 1934, *E. M. Bruce* 58 !

Distr. **U**?2, ?3 ; **K**4, 5 ; **T**2–3, 6–8 ; widely spread throughout tropical Africa and South Africa ; also in St. Helena and Ascension Is.

Hab. Edges of upland rain-forest and moist bamboo-thicket, secondary bushland, riverbanks in upland grassland ; 2400 (?–2500) m.

Syn. *R. pinnatus* Willd., *sensu stricto* ; DC., Prodr. 2 : 556 (1825) ; Focke in Bibl. Bot. 72 : 177 (1911) ; C. E. Gust. in Arkiv Bot. 26 (7) : 45 (1934) ; F.P.N.A. 1 : 241 (1948) ; T.T.C.L. : 480 (1949) ; F.C.B. 3 : 26 (1952)

? *R. kingaënsis* Engl. in E.J. 30 : 313 (1901) ; Focke in Bibl. Bot. 72 : 173 (1911) ; C. E. Gust. in Arkiv Bot. 26 (7) : 47 (1934) ; T.T.C.L. : 480 (1949). Type : Tanganyika, Livingstone Mts., Yawiri Mts., [S. of Bulongwa], *Goetze* 1192 (B, holo. †, LD, photo. !)
? *R. rungwensis* Engl. in E.J. 30 : 314 (1901) ; Focke in Bibl. Bot. 72 : 174 (1911) ; C. E. Gust. in Arkiv Bot. 26 (7) : 49 (1934) ; T.T.C.L. : 481 (1949). Type : Tanganyika, Rungwe Mt., *Goetze* 1164 (B, holo. †, LD, photo. !)

Note. There are certain inadequate specimens at Kew from Uganda (*Jarrett* 397A !, *Dummer* 3616 !) which may be this species. *R. kingaënsis* may, from a photograph of its holotype (LD !), be more correctly referable to the next variety.
See also under var. *afrotropicus* (below).

var. **afrotropicus** (*Engl.*) *C. E. Gust.* in Arkiv Bot. 26 (7) : 46 (1934) and in B.J.B.B. 13 : 273 (1935) ; F.P.N.A. 1 : 241 (1948) ; T.T.C.L. : 480 (1949) ; F.C.B. 3 : 27 (1952) ; F.W.T.A., ed. 2, 1 : 426 (1957). Types : many syntypes from several countries in tropical Africa, including Tanganyika (B, syn. †); Cameroons, *Preuss* (B, syn. †, LD, photo. !)

Prickles of flowering shoots up to 5(–6) mm. long. Inflorescence eglandular, pyramidal, up to 35 cm. long (or more) and to 25 cm. broad at the base, the basal branchlets up to 20 cm. long. Carpels tomentose or glabrous (sometimes tomentose only when young).

Uganda. Ruwenzori, Musandama, Dec. 1925, *Maitland* 1026 ! ; Mengo District : Kampala–Entebbe road, Kajansi Forest, Nov. 1935, *Chandler* 1471 ! & Kampala–Masaka road, Mar. 1936, *Michelmore* 1289 !
Kenya. Aberdare Mts., Chania Forest, Oct. 1943, *Logie in Bally* 7956 ! ; Teita Hills, Wusi, May 1931, *Napier* 1116 !
Tanganyika. Lushoto District : E. Usambara Mts., Amani, Nov. 1955, *Tanner* 2490 ! ; Ufipa District : Mbisi Mts., Nov. 1933, *Michelmore* 712 ! ; Morogoro District: Uluguru Mts., Oct. 1935, *Rounce* 507 !
Distr. **U**2, 4 ; **K**3, 4, 7 ; **T**2–4, 6, 8 ; Northern and Southern Rhodesia, Angola, Belgian Congo, Nigeria, Cameroons and Fernando Po, Liberia and French Guinea ; probably also in South Africa and St. Helena.
Hab. As for var. *pinnatus* ; 900–2400 m.

Syn. *R. pinnatus* Willd. subsp. *afrotropicus* Engl. in V.E. 3 (1) : 294 (1915)
R. pinnatus Willd. var. *defensus* C. E. Gust. in Arkiv Bot. 26 (7) : 46 (1934) from description and photograph. Type : Tanganyika, Bukoba District, near Itara, *Mildbraed* 148 (B, holo. † ; LD, photo. !)

Note. *R. pinnatus* is a variable species, and the two varieties described above are not always clearly distinguishable in terms of inflorescence form, which is perhaps the best character for distinction. Var. *afrotropicus* was described originally as a subspecies and it is true that in tropical Africa it appears largely to replace var. *pinnatus* which is more characteristic of S. Africa, thus the claim for subspecific grade is reasonable. On the other hand the available evidence is scarcely adequate to negative the possibility that the effects of habitat, coupled with the chances of collection, may not have presented a misleading picture, and for this reason I follow Gustafsson in adhering to varietal grade for each.

var. **subglandulosus** (*C. E. Gust.*) *R. Grah.* in K.B. 1957 : 406 (1958). Type : Kenya, Elgon, *Lugard* 509 (K, holo. !)

Stipitate glands, with or without gland-tipped pricklets, present on the calyx and pedicels. Carpels glabrous.

Kenya. Elgon, Jan. 1931, *Lugard* 509 !
Tanganyika. Arusha District : Mt. Meru, Nov. 1932, *Geilinger* 3885 ! ; Iringa District : Mufindi, 31 Oct. 1947, *Brenan & Greenway* 8254*
Distr. **K**3, ?5 ; **T**2, 6*, 7* ; not known elsewhere.
Hab. Probably as for var. *pinnatus* ; 1950–2200 m.

Syn. *R. pinnatus* Willd. forma *subglandulosus* C. E. Gust. in K.B. 1938 : 183 (1938)

Note. In Gustafsson's description the glands are described as subsessile, which is scarcely accurate as some of them are on stalks nearly 1 mm. in length. His " No tendency to produce glands has been observed before in *R. pinnatus* " refers no doubt only to stipitate glands.
Gland-tipped pricklets are absent in *Geilinger* 3885, but they occur in Gustafsson's holotype.

* This record added by Editor in proof.

16. **R. apetalus** *Poir.*, Encycl. 6 : 242 (1804) ; F.T.A. 2 : 374 (1871) ; Focke in Bibl. Bot. 72 : 176, fig. 72 (1911) ; C. E. Gust. in Arkiv Bot. 26 (7) : 53 (1934) and in K.B. 1938 : 180–1 (1938) ; F.C.B. 3 : 26 (1952) ; F.W.T.A., ed. 2, 1 : 426 (1958). Type : Réunion, *Commerson* (P, holo., K, photo. !)

A scrambling shrub. Flowering branches densely and softly white- or yellowish-pilose, sometimes villous, the hairs largely spreading ; prickles slender, hooked or straightly deflexed, up to 4–5 mm. long. Leaves imparipinnate (2–(3–)-jugate) or trifoliolate, or the uppermost simple ; leaflets rather variable in shape, commonly lanceolate, oblong-lanceolate, or oblong-elliptic, 6·5–9 × 4–6 cm., rather obtuse or the upper ones usually very acute, basally rounded to subcordate, sharply but shallowly serrate or biserrate with teeth ± 1(–2) mm. deep, dark green and variably hairy above (often rather evenly but thinly covered with appressed hairs), paler beneath, and softly pilose or more usually grey- or greenish- or whitish-tomentose, sometimes thickly covered with soft, whitish or ± fulvous hairs ; terminal petiolules 1–2(–3) cm. long ; the lateral ones up to 7 mm. long. Inflorescence paniculate, narrowly to broadly pyramidal, up to 21 × 9 cm., the basal branchlets spreading at a wide angle or ± horizontally ; flowers rather congested ; axis, branchlets and pedicels whitish-tomentose interspersed with variably numerous, fulvous or pale fulvous, stiff hairs. Petals O [or very small, (*fide* Focke)]. Calyx 5–8 mm. long, deeply divided into lanceolate, ± clasping lobes 4–7 mm. long with a protracted apex, hairy as the pedicels. Carpels varying from glabrous to densely hairy (and sometimes with a dense apical tuft). Fruit black.

Uganda. Kigezi District : base of Mt. Sabinio, *Eggeling* 1129 ! ; Ruwenzori, *Scott Elliot* 7912 ! ; Mengo District : Kipayo, Aug. 1914, *Dummer* 1008 !
Kenya. Ravine District : Eldama Ravine, *Whyte* ! ; Aberdare Mts., Chania Forest, Oct. 1943, *Logie in Bally* 7958 ! ; Kericho District : Sotik, June 1953, *Verdcourt* 951 !
Tanganyika. Bukoba District : Bugere, May 1948, *Ford* 531 ! ; Moshi District : Lyamungu, Aug. 1932, *Greenway* 3115 ! ; Ufipa District : Mbisi, Apr. 1950, *Bullock* 2812 !
Distr. **U**2–4 ; **K**3–5 ; **T**1–4, ?5 ; Ethiopia, Nyasaland, Belgian Congo, British Cameroons ; also in the Mascarene Islands and possibly in Nigeria.
Hab. Edges of upland rain-forest, upland bushlands, thickets, roadsides ; 1275–2100 m.

Syn. *R. exsuccus* A. Rich., Tent. Fl. Abyss. 1 : 256 (1848) ; Focke in Bibl. Bot. 72 : 176 (1911) ; C. E. Gust. in Arkiv Bot. 26 (7) : 36 (1934) and in K.B. 1938 : 181 (1938) ; F.W.T.A., ed. 2, 1 : 426 (1958). Type : Ethiopia, Semen, *Schimper* 867 (BM, K, iso. !)
R. interjungens C. E. Gust. in Arkiv Bot. 26 (7) : 32, t. 5 (1934) and in B.J.B.B. 13 : 269 (1935) and in K.B. 1938 : 182 (1938) ; F.P.N.A. 1 : 239 (1948) ; T.T.C.L. : 480 (1949) ; F.C.B. 3 : 24 (1952). Type : Kilimanjaro, Namui, *Endlich* 388 (B, holo. †, LD, photo. !)
R. apetalus Poir. forma *pyramidalis* C.E. Gust. in B.J.B.B. 13 : 274 (1935) ; F.P.N.A. 1 : 241 (1948) ; F.C.B. 3 : 26 (1952). Types : Belgian Congo, Ruwenzori, Lamia Valley, *Bequaert* 4241 & 4241 bis (BR, syn. !)

Note. A variable bramble, not always easily separated from *R. rigidus* or *R. pinnatus* (*sensu lato*).
Despite the distinctions given by Gustafsson for *R. exsuccus*, I follow Oliver and Focke in regarding its affinity to *R. apetalus* too close to warrant separation.

17. **R. adolfi-friedericii** *Engl.* in Z.A.E. : 223 (1911) ; C. E. Gust. in Arkiv Bot. 26 (7) : 35 (1934) and in K.B. 1938 : 181 (1938) ; T.T.C.L. : 479 (1949) ; F.C.B. 3 : 25 (1952). Type : Ruanda-Urundi, Rugege Forest, Rukarara [Lukarara] R., *Mildbraed* 706 (B, holo. †, LD, photo. !)

A scrambling shrub (? or simple, erect canes), 1·5–2 m. tall (? or more). Flowering branches (and, apparently the turions) densely covered with ± fulvous, ± bristly hairs ; prickles up to 6 mm. long, ± deflexed, narrow.

Leaves imparipinnate (2-jugate), or the uppermost trifoliolate; leaflets ovate to ovate-elliptic, or oblong, 4–8 × 3–5 cm., acute, sometimes long-acuminate, basally round to subcordate, shallowly but sharply serrated or biserrated with excurrent nerves, green and rather thinly hairy above, softly whitish-tomentose beneath; terminal petiolule 1–3 cm. long, the lateral ones not exceeding 5 mm. Inflorescence a short, congested, narrowly cylindrical, terminal panicle scarcely exceeding 8 × 4 cm., not much exceeding the leaves, with the lowest branches 2·5–6 cm. long, with or without axillary panicles below. Pedicels hairy, 3–4 mm. long. Calyx ± 6·5 mm. long, deeply divided into lanceolate or ovate-lanceolate, acuminate lobes ± 5·75 mm. long, whitish-tomentose on the inside, interspersed with pale fulvous bristly hairs on the outside. Petals absent, or fugacious. Carpels glabrous. Fruit yellowish when young, becoming red or orange-red later, edible (see Note).

UGANDA. Kigezi District: Kachwekano Farm, Sept. 1949, *Purseglove* 3099! & Dec. 1949, *Purseglove* 3163!; Masaka District: Lake Nabugabo, July 1937, *Chandler* 1768!
KENYA. Kiambu District: Kabete, Aug. 1947, *Bogdan* 1080!
TANGANYIKA. Bukoba District: June 1936, *Culwick* 4!; Moshi District: N. slope of Kilimanjaro, continuation of Laitokitok Ridge, Nov. 1932, *C. G. Rogers* 122!; Njombe, Dec. 1931, *Lynes* F.j. 21!
DISTR. **U**2, 4; **K**?3, 4; **T**1, 2, 7; Belgian Congo, Ethiopia.
HAB. Edges of upland rain-forest, secondary bushland and thickets; 1140–2100 m.

SYN. *R. adolfi-friedericii* Engl. var. *rubristylus* C. E. Gust. in Arkiv Bot. 26 (7): 35 (1934) and in B.J.B.B. 13: 269 (1935); F.P.N.A. 1: 239 (1948); F.C.B. 3: 25 (1952). Types: Kenya, Mt. Kenya, *Fries* 730, 933, 1802 (U, syn.!)

NOTE. This bramble differs from the closely allied *R. apetalus* by its dense, short panicle, and its red fruit (which are however recorded as black when fully ripe in *Chandler* 1768). The pedicels are certainly short, and, as stated by Engler, apparently shorter than those of *R. apetalus*, but field studies are needed to decide whether it should be regarded as specifically distinct.
The only turion-leaf seen has broadly ovate to ovate-elliptic, 2-jugate leaflets.

18. **R. niveus** *Thunb.*, Dissert. Rubi: 9 (1813); Focke in Bibl. Bot. 72: 182 (1911). Type uncertain

An erect, bushy shrub. Stems glabrescent. Leaves imparipinnate (2–3-jugate); leaflets green and thinly hairy above, whitish-grey-tomentose beneath; the terminal leaflets broadly ovate, basally cordate, larger than the elliptic or ± ovate-elliptic lateral leaflets. Inflorescence short, ± corymbose, leafy and usually overtopped by leaves. Calyx lobes ± 5 mm. long, exceeding the petals, densely pilose. Carpels densely tomentose.

KENYA. Nairobi, July 1937, *van Someren* 142!
TANGANYIKA. Moshi District: Lyamungu, Mar. 1944, *Wallace* 1194!; Lushoto, July 1955, *Benedicto* 47!
DISTR. **K**4; **T**2, 3; an Indian and Malayan bramble, introduced into East Africa.

SYN. *R. lasiocarpus* Sm. in Rees, Cycl. 30, No. 6 (1819); Fl. Brit. Ind. 2: 339 (1878) *max. ex parte fide* Focke. Type: India, *Rottler* (LINN–SM, holo.!)

4. CLIFFORTIA

L., Sp. Pl.: 1038 (1753) & Gen. Pl. 5: 460 (1754)

Procumbent or erect, much branched, dioecious or monoecious shrubs or more rarely trees. Leaves of tropical East African species trifoliolate, ericoid, fasciculate. Petals 0. Male flowers with 3–4 petaloid calyx-lobes which may be free or connate ± to the centre. Stamens 3–50. Female flowers with calyx-lobes connate below into an urceolate tube which wholly envelopes

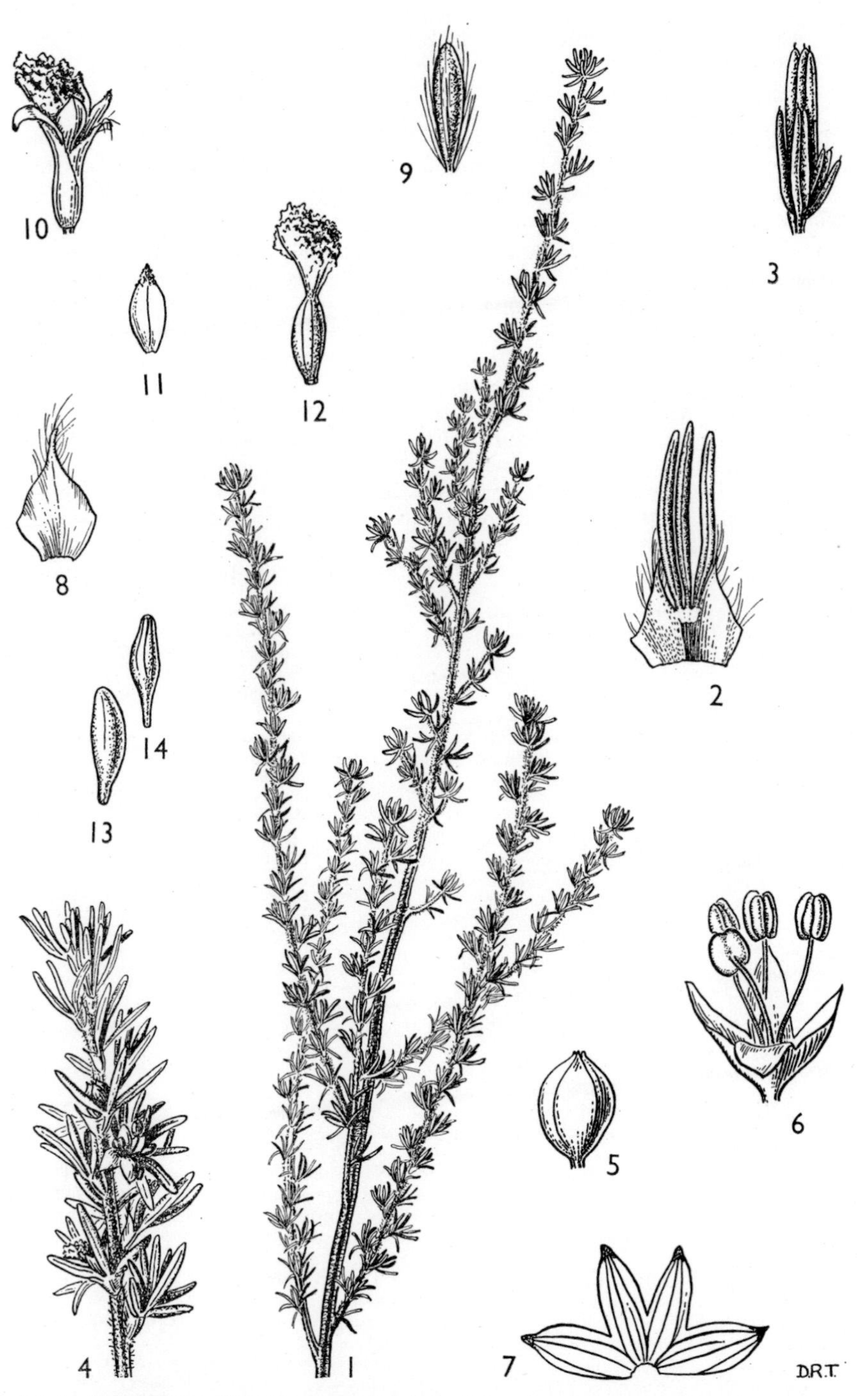

FIG. 4. *CLIFFORTIA NITIDULA*, from *Dale* 3432,—**1**, leafy branch, to show habit, × ½; **2**, leaf, with stipules, × 4; **3**, young leaves, to show immature habit, × 4; **4**, apex of flowering branch, × 2; **5**, ♂ flower-bud, × 4; **6**, ♂ flower, lateral view, × 4; **7**, calyx (of ♂ flower) opened out, × 4; **8**, bracteole (of ♀ flower), × 6; **9**, ♀ flower-bud, × 6; **10**, ♀ flower, lateral view, × 6; **11**, calyx-lobe of ♀ flower, × 6; **12**, pistil, × 6; **13**, fruit, × 5; **14**, achene, × 5.

the receptacle. Achenes 1(–2), superior, cylindrical, closely enveloped by the receptacle-wall ; style filiform ; stigma with a feathery margin.

A genus confined, except for the following species, to South Africa.

C. nitidula *R. E. & T. C. E. Fries* in N.B.G.B. 8 : 649 (1923) ; H. Weim., Monogr. Cliff. : 47 (1934) ; Brenan in Mem. N. Y. Bot. Gard. 8 : 432 (1954). Types : Tanganyika, Uluguru Mts., Lukwangule, *Stuhlmann* 9160 & *Goetze* 257 (B, syn. †)

A much-branched shrub, 4–5 m. high, apparently monoecious. Branches reddish-brown, densely grey-villous when young, the outer bark tending to peel away. Leaflets 4–7 mm. long, narrowly linear-oblanceolate to narrowly linear due to the degree of revolution of the margins which usually hide the midrib, subacute, leathery, glabrous, shining above, deep green, sometimes red-tipped ; petioles 1–2·5 mm. long, brown. Stipules triangular, acute, pilose on the inside. Flowers bracteate, sessile, borne in the leaf-fascicles. Male flowers with broadly oblong sepals (3–)3·8–4(–4·5) mm. long, apically narrowed and mucronate. Stamens 4 ; filaments reddish-purple or violet ; anthers reniform, violet. Female flowers greenish-white. Receptacle oblong, becoming elongated, with 2 narrow, longitudinal wings. Sepals oblanceolate, ± 1·75 mm. long, ± apiculate. Stigma purplish ; style very short ; achene solitary, 3 mm. long, asymmetrically cylindrical. Fig. 4, p. 41.

KENYA. Elgeyo District : Marakwet Hills, June 1935, *Dale* 3432 ! ; Aberdare Mts., July 1948, *Hedberg* 1503 ! ; Mt. Kenya, R. Kongoni, 13 Feb. 1922, *Fries* 1575

TANGANYIKA. Morogoro District: Nguru Mts., Mt. Mesumba, 20 July 1933, *Schlieben* 4190 ! ; Rungwe District : NW. of Rungwe Mt., by the Lower Kiwira River, Oct. 1947, *Brenan & Greenway* 8194 ! ; Njombe District : Elton Plateau, R. Lumakanya Oct. 1954, *Willan* 184 !

DISTR. **K**3, 4 ; **T**6, 7 ; Northern and Southern Rhodesia, Nyasaland, Angola.

HAB. Moist bamboo-thicket, upland moor and riverbanks in upland grassland 2040–2950 (?–3150) m.

SYN. [*C. linearifolia* sensu F.T.A. 2 : 379 (1871), *non* Eckl. & Zeyh.]

C. linearifolia Eckl. & Zeyh. var. *nitidula* Engl. in E.J. 26 : 376 (1899). Types : Angola, Huila, Humpata, *De Melho Ramalho* (B†, COI, syn.) & Tanganyika, Uluguru Mts., Lukwangule, *Stuhlmann* 9160 (B, syn.†)

C. aequatorialis R. E. & T. C. E. Fries in N.B.G.B. 8: 649 (1923) ; H.Weim., Monogr. Cliff. : 50 (1934). Type : Kenya, Aberdare Mts., *Fries* 2438 (U, holo., K !, S, iso.)

C. nitidula R. E. & T. C. E. Fries subsp. *angolensis* H. Weim., Monogr. Cliff. : 49 (1934). Types : Angola, Huila, *Welwitsch* 1277 (C, COI, K !, syn.) & Huila, Humpata, *De Melho Ramalho* (B†, COI, syn.) & Bié, R. Longa, *Baum* 650 (B†, COI, G, HBG, K !, S, W, Z, syn.) & Bié, R. Cuito, *Gossweiler* 2566 (COI, K !, syn.)

C. nitidula R. E. & T. C. E. Fries var. *aequatorialis* (R. E. & T. C. E. Fries) Brenan in Mem. N. Y. Bot. Gard. 8 : 433 (1954)

C. nitidula R. E. & T. C. E. Fries var. *angolensis* (H. Weim.) Brenan in Mem. N. Y. Bot. Gard. 8 : 433 (1954)

NOTE. I follow Brenan in regarding *C. aequatorialis* and *C. nitidula* subsp. *angolensis* as variants of *C. nitidula* which have been separated only on leaf-characters which are not constant in respect of any of these three taxa, there being intervariation from one to the other. For this reason it is proposed to relegate the first two to synonymy with the last-named.

It is at least probable that *C. nitidula* should be considered as a subspecies of the South African and Southern Rhodesian *C. linearifolia* Eckl. & Zeyh. But the leaves of the latter have a markedly thicker midrib which is not usually obscured by the revolution of the margins.

In using the epithet *nitidula*, the authors evidently intended their species to be based on Engler's variety of *C. linearifolia*. However, as they used partly different types, the latter's name has been omitted from the authority citation.

5. HAGENIA

J. F. Gmel., Syst. 2 : 613 (1791)
[*Banksia* (*Bankesia*) Bruce, Trav. 5 : 73 (1790), *non* Linn. f. (1781), *nom. illegit.*]

Dioecious or polygamous tree. Leaves imparipinnate. Flowers small, anemophilous, aggregated into large many-flowered panicles. Petals 0 or rudimentary. Calyx adnate to the receptacle, apically produced into (4–)5 lobes (" bracteoles "). Receptacle closed at the throat, terminating in (4–)5 (inner) lobes which usually alternate with the calyx-lobes : thus each flower bears an " inner " and an " outer " row of calyx-lobes, all membranous. Male flowers with the outer calyx-lobes smaller than the inner, the receptacle not accrescent, the ovaries aborting ; stamens 12–20. Female flowers with the outer calyx-lobes and receptacle accrescent, the outer lobes much larger than the inner and serving ultimately to parachute the ripe fruit by rotation ; stamens with very short filaments and abortive anthers ; styles 2 ; carpels 2, free within the receptacle ; ovule single, pendant. Hermaphrodite flowers apparently rare, similar to ♀ flowers but having (at least some) fertile anthers. Achenes brown, (often only 1 developing), reticulately rugose, ± ovoid with a thin, fragile pericarp.

A monotypic genus, confined to Africa. A popular and highly effective anthelmintic.

H. abyssinica (*Bruce*) *J. F. Gmel.*, Syst. 2 : 613 (1791) ; F.P.N.A. 1 : 254 (1948) ; T.T.C.L. : 475 (1949) ; F.C.B. 3 : 16 (1952) ; I.T.U., ed. 2 : 332, photo. 53 (1952). Type : Ethiopia

A rather slender tree up to 20 m. tall, with brown or reddish-brown, readily strip-peeling bark. Young twigs densely villous with yellow-brown soft ascending hairs 3–4 mm. long, with villous ring-scars of previous petiole-bases. Leaves petiolate, viscid, up to 40 cm. long : main leaflets commonly 6–8 on each side, sessile or almost so, narrowly oblong, commonly 12–15 × 3·5–5·2 cm., acuminate, basally rounded to subcordate, finely serrated and villous on the margins, with the primary nerves prominent beneath and the intervening veins reticulate, dark green, usually pubescent above, lighter green, densely villous with soft silvery hairs varying to glabrous beneath ; rhachis usually with very small, ± orbicular secondary leaflets, 2–10 mm. long, between the main ones. Petioles 12–13 cm. long, winged due to the stipules being adnate for almost their whole length. Inflorescence a handsome, much-branched terminal drooping panicle 30–60 × 20–30 cm. at fullest maturity (the ♂ panicles usually relatively narrower), the branches villous, subtended by membranous and caducous leafy bracts ; internodes ± zig-zag ; pedicels viscid ; flowers subtended by 2(–3) broadly rounded bracts. Male flowers orange-buff to white, ± 7·5 mm. diameter at anthesis ; calyx-tube 2–3 mm. long ; outer calyx-lobes oblong or ovate to obovate, 1–1·5 × 0·5–0·75 mm., the inner lobes larger, 4–6 × 2·5–3 mm., reflexing and concave on the under side. Female flowers rather viscid, more red, up to 1·75 cm. in diameter ; outer calyx-lobes unequal, oblong-linear varying to obovate, 7 × 3·75 to 9 × 5 mm., clearly veined ; the inner lobes smaller, broadly ovate, 2·5 × 2 mm. Styles 1·25–1·5 mm. long ; stigmas capitate ; carpels asymmetrically ovoid, apically villous. Fig. 5, p. 44.

UGANDA. Acholi District : Imatong Mts., Apr. 1938, *Eggeling* 3536 ! ; Kigezi District : Luhiza, June 1948, *Purseglove* 2704 ! ; Mbale District : Elgon, Bulambuli, Nov. 1933, *Tothill* 2297 !

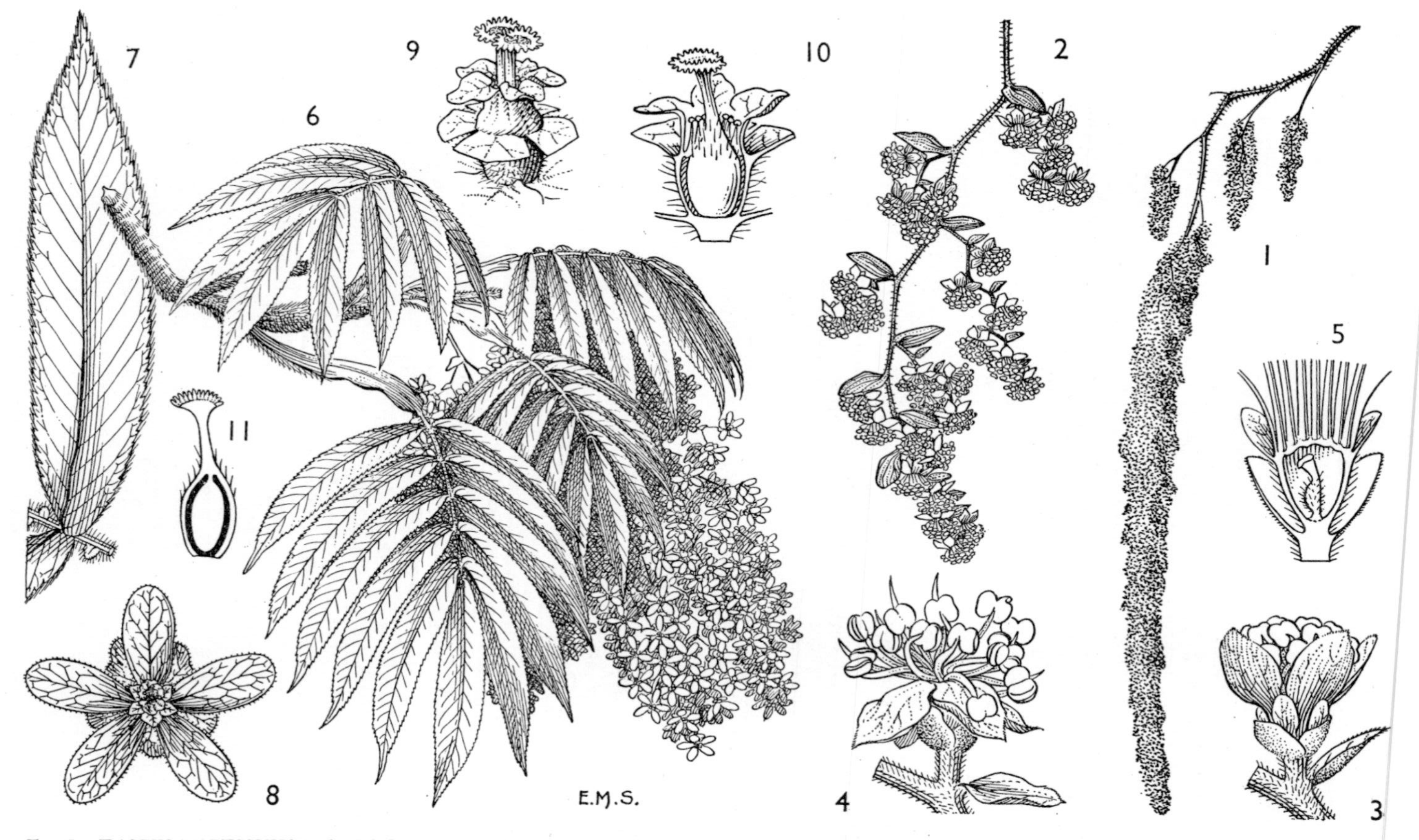

FIG. 5. *HAGENIA ABYSSINICA*—**1,** ♂ inflorescence, × $\frac{1}{6}$; **2,** part of same, × 1; **3,** ♂ flower, partly open, × 4; **4,** ♂ flower, × 4; **5,** part of ♂ flower, in L.S. to show pistillode, × 8; **6,** branch with ♀ inflorescence, × $\frac{1}{3}$; **7,** leaflet, × 1; **8,** ♀ flower, dorsal view, × 2; **9,** ♀ flower, lateral view, with outer calyx-lobes removed, × 4; **10,** ♀ flower partly in L.S., to show staminodes and pistil, × 4; **11,** pistil, × 4. 1–5, from *Tothill* 2365; 6–11, from *Richards* 6666.

KENYA. Elgeyo District : Cherangani Mts., Sept. 1949, *Maas Geesteranus* 6324 !; Kericho District : SW. Mau Forest Reserve, Aug. 1949, *Maas Geesteranus* 5639 !; Mt. Kenya, Urumandi, Aug. 1944, *Le Pelley in Bally* 3483 !
TANGANYIKA. Arusha District : west side of Mt. Meru, Sept. 1932, *B. D. Burtt* 4092 ! ; Lushoto District : W. Usambara Mts., Gologolo, May 1922, *A. S. Adamson* 69 ! ; Mbeya District : Poroto Mts., Wentzel Hekmann Crater, Aug. 1936, *B. D. Burtt* 6231 !
DISTR. **U**1–3 ; **K**3–5 ; **T**2, 3, 7 ; the Belgian Congo, Ethiopia, Sudan (Imatong Mts.), and also Northern Rhodesia and Nyasaland (Nyika Plateau).
HAB. Upland rain-forests, often above the moist bamboo-thickets, and in upland evergreen bushland ; 2400–3600 m.

SYN. *Banksia abyssinica* Bruce, Trav. 5 : 73 (1790)
Brayera anthelmintica Kunth. in Brayer, Not. Vermif. (1824) ; DC., Prodr. 2 : 588 (1825) ; F.T.A. 2 : 380 (1871) ; T.S.K. : 58 (1936). Type : uncertain
B. anthelmintica Kunth. var. *psilanthera* Bitter in F.R. 12 : 378 (1913). Type : Ethiopia, Tigré, Agame, *Schimper* 659 (BM, isolecto. !)
B. anthelmintica Kunth. var. *epirhagadotricha* Bitter in F.R. 12 : 378 (1913), *ex descr.* Type : Kilimanjaro, between Marangu and Machame, *Meyer* 321 (B, holo.†)
Hagenia abyssinica (Bruce) Gmel. var. *viridifolia* Hauman in B.J.B.B. 22 : 90 (1952) ; F.C.B. 3 : 17 (1952). Type : Belgian Congo, Mukule, *Bequaert* 6299 (BR, holo.)

NOTE. Almost all herbarium specimens consulted were either ♂ or ♀ trees ; in two cases only was polygamy (♂ and ☿) evident, and in one of these the ☿ flowers had apparently fertile stamens but abortive ovaries, thus the example, although in essence ☿ was in fact functionally only ♂.
Var. *viridifolia* is a form with glabrous or nearly glabrous leaf-surfaces. Typical *H. abyssinica* is assumed to have softly pubescent to villous leaves, but between this and the variety there is every intermediate in this character. As there is no clear geographical or other character correlatable with hairiness, it is proposed to sink the variety into synonymy with the species.

6. PYGEUM

Gaertn., Fruct. 1 : 218 (1788)

Hermaphrodite or occasionally polygamous trees or shrubs. Leaves simple, entire or serrate, petiolate. Calyx cup-like, with 5 lobes later deciduous leaving a narrow circular rim. Petals (0–)5(–12), small, inserted in the mouth of the calyx-tube. Stamens (10–)21–33(–35), perigynous. Ovary superior, sessile, narrowed above ; ovules 2, paired ; style terminal. Fruit drupaceous, 1–2-seeded ; pericarp dry.

A genus of some 75 species, of the warmer parts of Asia, with one (or perhaps two) endemic species in Africa.

P. africanum *Hook. f.* in J.L.S. 7 : 191 (1864) ; F.T.A. 2 : 373 (1871) ; T.S.K. : 57 (1936) ; F.P.N.A. 1 : 256 (1948) ; T.T.C.L. : 477 (1949) ; F.C.B. 3 : 32 (1952) ; I.T.U., ed. 2 : 335 (1952) ; F.W.T.A., ed. 2, 1 : 426 (1958). Type : British Cameroons, Cameroon Mt., *Mann* 1207 (K, lecto. !)

A fine evergreen tree with somewhat pendulous branches, up to 36 m. tall (rarely a shrub to 4·5 m. tall), glabrous except for the flowers. Bark dark brown to grey, longitudinally fissured ; branchlets smooth with lenticels 1–2 mm. long, their rims eminently labial. Leaves matt or shining, thick and leathery, elliptic-oblong, ovate, or obovate, commonly three times as long as broad and up to 15 × 5·2 cm., acute or obtuse or shortly acuminate, basally shortly cuneate to rounded, crenate-serrate with many shallow serratures each with a black, readily caducous, terminal cusp. Stipules ± linear, 1·5–2·0 mm. long, caducous. Inflorescence a congested or open, erect, spreading or pendulous, sometimes rather fleshy raceme 3·5–8·0 cm. long, arising

FIG. 6. *PYGEUM AFRICANUM*—**1,** leaf, × 1; **2,** apex of flowering branch, × 1; **3,** flower, × 4; **4,** flower, from beneath, showing sepals, petals and lower part of filaments, × 4; **5,** petal, × 8; **6,** part of flower in L.S. to show pistil and ovule, × 4; **7,** fruiting raceme, × 1; **8,** fruit, × 2. 1, from *Parry* 70; 2, from *Moon* 765; 3–6, from *Milne-Redhead & Taylor* 10877; 7–8, from *Bally* 4356.

singly or in fascicles of 2–4 from axils on the new growth of the preceding year, and sometimes branched from near the base, 7–15-flowered ; pedicels (3–)5–7(–11) mm. long, usually ± rigid. Calyx-tube 3·5–5·5 mm. in diameter at the mouth, glabrous and often rather fleshy outside ; inner receptacle surface often villous ; calyx-lobes deltoid, 1·5 mm. long, acute, glabrous or with scattered hairs. Petals 5, creamy-white, spathulate, ± 2 mm. long, ciliate to villous, often revolute. Filaments (1·5–)2 mm. long. Style 2 mm. long, usually with at least a few hairs and often villous ; stigma 2- or obscurely 3-lobed ; ovary ± ovoid, often villous. Drupe red to dark reddish-brown when ripe, rounded-ellipsoid, up to 0·7 cm. long, 1·1 cm. in diameter, and appearing as if bilocular, the hardened remains of the style persistent as a cusp. Fig. 6.

Uganda. Acholi District : SE. Imatong Mts., Lomwaga Mt., Apr. 1945, *Greenway & Hummel* 7283 ! ; Kigezi District : Amahingo, Aug. 1949, *Purseglove* 3054 ! ; Busoga District : Mutai, 1918, *Fyffe* 30 !

Kenya. Northern Frontier Province : Mt. Nyiro, Feb. 1947, *Mrs. J. Adamson in Bally* 6159 ! ; Elgon, Dec. 1930, *Lugard* 445 ! ; Machakos District : N. side of Chyulu Hills, Apr. 1938, *Bally* 20 *in C.M.* 7653 !

Tanganyika. South slopes of Ngorongoro Crater, Sept. 1932, *B. D. Burtt* 4325 ! ; Buha District : Murungu, Aug. 1950, *Bullock* 3208 ! ; NE. of Mpwapwa, Aug. 1930, *Greenway* 2473 !

Distr. **U**1–4 ; **K**1, 3–6 ; **T**1–5, 7, 8 ; Cameroons, Fernando Po, Angola, Natal, and the Cape Province.

Hab. Upland rain-forest and riverine forest, or on termite-hills in *Brachystegia* woodland ; 900–3000 m.

Syn. *P. crassifolium* Hauman in B.J.B.B. 22 : 93 (1952) ; F.C.B. 3 : 33 (1952) ; *saltem parte*

Note. A useful timber-tree with an unpleasantly smelling wood, locally known as Red Stinkwood ; also used for medical purposes, and for poison for arrows.

The position regarding *P. crassifolium* is not very satisfactory. *Stolz* 2279 ! and 2414 ! (from Tanganyika, Rungwe District, Ukinga Mts., Mwakaleli—the former being cited as *P. crassifolium* in the type description)—both fall within the variational range of *P. africanum*, from which it is scarcely reasonable to separate them. The same may be said of *Humbert* 8040 ! and *Chapin* 416 !, both from the Belgian Congo, and cited as *P. crassifolium* in F.C.B. (see above). But Bequaert's specimens from Ruwenzori (1912 and 1914—the former unlocalized and the latter from the Lanuri Valley) which Haumen has labelled " Typus " have a rather markedly different appearance, and may be taxonomically separable from *P. africanum*, or—more likely—may represent a form of this species indicating an extreme in fleshiness. Further collecting in the Lanuri Valley, at 3000 m. is required.

7. CHRYSOBALANUS

L., Sp. Pl. : 513 (1753) & Gen. Pl., ed. 5 : 229 (1754)

Trees or shrubs. Leaves simple, petiolate, entire ; blade with two small basal glands. Flowers with a shallow receptacle. Calyx-lobes and petals 5. Stamens 15 or more, all or partly fertile, the filaments usually partly united towards the base, inserted all round the rim of the receptacle. Carpel single, sessile, inserted at the base of the receptacle. Fruit drupaceous, with a large stone.

A tropical, Afro-American and largely maritime genus of few species, three occurring in the Belgian Congo.

C. stuhlmannii *Dammer* in Engl., P.O.A. C : 191 (1895). Type : Tanganyika, Chaya Lake, *Stuhlmann* 430 (B, holo.†)

A tree with young branchlets, petioles and inflorescences densely covered with rusty, silky hairs. Leaves leathery, very shortly petiolate, oblong, apically obtuse and basally slightly emarginate ; midrib and nerves rusty-pilose, otherwise glabrous and shining on the upper surface. Inflorescence (immature) paniculate. Calyx-lobes ovate-lanceolate ; receptacle cup-like. Petals broadly obovate. Stamens 15. Style basal.

TANGANYIKA. Dodoma District : Chaya Lake, *Stuhlmann* 430
DISTR. T5 ; known only from a single gathering.
HAB. Unknown.

NOTE. There is doubt as to the identity of the above-described tree of which apparently no specimens exist and which despite its description, may more correctly belong to another genus. Further collecting in the area of Chaya Lake is however most desirable.

8. PARINARI

Aubl., Pl. Guiane Fr. 1 : 514 (1775)

Parinarium Juss., Gen.: 342 (1789)

Trees or shrubs, sometimes low and rhizomatous. Leaves simple, entire, with 2 glands on the upper side of the petiole or at the junction of the petiole with the blade. Bracts, bracteoles, and stipules caducous. Flowers hermaphrodite. Calyx-tube turbinate and often obliquely bent (subgen. *Sarcostegia*) or cup-shaped and unilaterally gibbous or ± so (subgen. *Parinari*); calyx-lobes 5, triangular-acute (*Parinari*) or ± rounded, concave (*Sarcostegia*). Petals 5, caducous. Fertile stamens very many (20–40), long-exserted (*Sarcostegia*) or few, 7–8, not or scarcely exserted (*Parinari*), confined to the dorsal and lateral rims of the strongly concave receptacle, and replaced at the frontal rim by very short staminodes. Ovary dorsifixed ; carpels usually 1, rarely 2–3, each bilocular (sometimes only partially so) ; ovules 1 to each cell ; style basal, long-exserted (*Sarcostegia*) or short, included or ± so (*Parinari*). Fruit drupaceous, ellipsoid, obovoid or ± globose, 1-seeded.

A tropical American, Asian and African genus of some important hardwood timber trees.

Two subgenera are represented in the area covered by this Flora, namely :—
Parinari . . . . spp. 1–3
Sarcostegia (Benth.) Hauman . . . . spp. 4, 5

Dwarf rhizomatous shrub with caespitose shoots seldom exceeding 60 cm. in height, and with very long roots . . . . 1. *P. capensis*
Trees or shrubs, but not dwarf and rhizomatous:
Leaves with 15–24(–32) " pairs " of primary nerves ; calyx (from joint in pedicels to base of lobes) not exceeding 4–5 mm. in length, cup-shaped, basally gibbous or ± so ; calyx-lobes acute :
Leaves acuminate, or ± evenly narrowed to an acute apex, commonly 1·5–2·5 cm. broad . 2. *P. excelsa*
Leaves terminally rounded or broadly obtuse, commonly 2·5–4·5 cm. broad . . . . 3. *P. curatellifolia*
Leaves with 6–8(–10) " pairs " of primary nerves ; calyx (measured as above) 5–17·5 mm. long, turbinate, sometimes bent, not or scarcely gibbous ; calyx-lobes rounded or ± so :
Leaves terminally acuminate or narrowed to a ± acute apex ; calyx (measured as above) 5 mm. long . . . . 4. *P. goetzeniana*
Leaves terminally rounded or very obtuse ; calyx (measured as above) 7–15(–17·5) mm. long 5. *P. polyandra*

1. **P. capensis** *Harv.* in Fl. Cap. 2 : 596 (1862) ; R. Grah. in K.B. 1957 : 230 (1957). Types : South Africa, unlocalized, *Zeyher* 537 (K, syn.!) ; Aaapjes River, *Burke* 518 (K, syn.!)

A dwarf shrub, with very long branched rhizomes, and erect caespitose branched stems, often very short but up to 75 cm. tall, covered above with yellow or silvery hairs, glabrescent below. Leaves shortly petiolate, narrowly oblanceolate to oblong or oblong-elliptic, 4–11·5 × 1·5–5·8 cm., rounded (sometimes emarginate) to broadly obtuse, basally cuneate to rounded or cordate, glabrous and green above, thickly white-tomentose beneath ; primary nerves 15–30 " pairs," obvious on both surfaces. Inflorescence terminal, or axillary to the uppermost 1 or 2 leaves, paniculate, branched ; panicles usually dense and shaggy with long ± spreading silvery or brown hairs ; cymes commonly triflorous. Flowers 4–6·5 mm. long (from base of calyx). Calyx cup-shaped, ± abruptly and unilaterally gibbous, the lobes (1·5–)1·75–2(–3) mm. long, triangular. Filaments of fertile stamens 1·5–1·75(–2) mm. long ; staminodes 0·5–1·0 mm. long. Style 3–3·5 mm. long, shaggy towards the base. Drupe subglobular or obovoid-globular, yellow when mature, but black and liberally spotted with light brown patches of cork when dried.

subsp. **latifolia** (*Oliv.*) *R. Grah.* in K.B. 1957 : 230 (1957). Type : Angola, Cuanza Norte, Cazengo, *Welwitsch* 1286 (K, iso.!)

Stems commonly 30–75 cm. tall, not gnarled, bearing leaves throughout. Leaves oblong-elliptic or oblong, varying to ovate, up to 5·8 cm. broad, commonly 7–9 × 2·5–5·0 cm., usually rounded to cordate at the base.

TANGANYIKA. Kigoma District : about 57·5 km. S. of Uvinza, Aug. 1950, *Bullock* 3264!

DISTR. **T4** ; Angola, the Belgian and French Congo, Northern Rhodesia, Portuguese East Africa.

HAB. *Brachystegia* woodland, and associated grasslands, often on sand ; 1700 m.

SYN. *P. capensis* Harv. var. *latifolia* Oliv. in F.T.A. 2 : 369 (1871)
P. curatellifolia [Planch. ex] Benth. var. *fruticulosa* R. E. Fries, Schwed. Rhod.-Kongo-Exp. 1 : 60 (1914). Types : Northern Rhodesia, Kamindas, *Fries* 646A and 659 (U, syn.)
P. pumila Mildbr. in Wiss. Erg. Zweit. Deutsch. Zentr.-Afr.-Exp. 1910–11, 2 : 4–5 (1922) ; Hauman in B.J.B.B. 21 : 193 (1951) and in F.C.B. 3 : 67 (1952). Type : uncertain
P. latifolia (Oliv.) Exell in J.B. 66, suppl. : 160 (1928)

NOTE. Subsp. *capensis*, which replaces subsp. *latifolia* in South Africa but which does not extend north of Angola and Northern Rhodesia, differs in being only 1·5–6(–15) cm. tall, in having tufts of leaves and flowers only at the apices of the very short, gnarled stems, and in its narrower, oblanceolate to elliptic-lanceolate leaves, up to 9 × 2 cm. (commonly 4–6·5 × 1·1–1·5 cm.), usually narrowed to the base. The two subspp. are clearly definable as extremes, but intermediates occur. Annual burning, or the absence of it, may play an important part in their habit.

2. **P. excelsa** *Sabine* in Trans. Hort. Soc. 5 : 451 (1824) ; R. Grah. in K.B. 1957 : 229 (1957). Type : Sierra Leone, *G. Don* (K, lecto.!)

A large evergreen tree up to 45 m. tall, with a thick rounded or flatly-spreading crown ; with or without basal buttresses up to 3 m. high. Bark grey-brown, rough, finely and often long-fissured. Leaves petiolate, lanceolate, elliptic, ovate-elliptic, lanceolate- or oblanceolate-elliptic, 5·5–12 (–15) × 1·5–3·5(–5) cm., long, narrowed to an acute or obtuse apex, sometimes acuminate, similarly or more shortly narrowed to the base (or there rounded to cordate on sterile branches) ; primary nerves up to 24 " pairs," irregularly parallel, very prominent below ; dark green and glabrous and sometimes shining above, buff and usually with ± appressed fulvous hairs on the midrib below and sometimes on the primary nerves, with the intervening tissues variably glabrous to densely hirsute with appressed hairs. Inflorescence paniculate, terminal or axillary, much-branched, shortly but thickly hairy to shaggy with irregularly spreading or ascending silky or brown hairs. Flowers in (usually) triflorous cymes, 4–8 mm. long. Calyx cup-shaped, basally abruptly gibbous, the lobes triangular, 1·75–3·5 mm. long, acute, rather spreading, hairy as the bracts. Petals white, ovate, readily caducous, of approximately the same length as the calyx-lobes. Filaments of antheriferous stamens 1·75–3 mm. long ; staminodes reduced to very small teeth not exceeding 0·75 to 2 mm. long. Ovary shaggy with long hairs ; style 2·5–5 mm. long, the stigma flatly capitate. Drupe ovoid to obliquely ellipsoid, up to 5 × 3·2 cm. when mature and dried, sometimes suborbicular, covered with small, thin patches of cork.

subsp. **holstii** (*Engl.*) *R. Grah.* in K.B. 1957 : 229 (1957). Type : Tanganyika, Lushoto District, Mlalo, *Holst* 2425 (K, isolecto. !)

Indumentum of the inflorescence brown or fulvous. Flowers 4–5 mm. long. Calyx-lobes 1·75–2(–2·5) mm. long. Filaments (1·75–)2–2·5 mm. long ; staminodes not more than 0·75 mm. long. Style (2·5–)3–3·5 mm. long.

UGANDA. Bunyoro District : Budongo Forest, June 1932, *Harris in Brasnett* 841 ! ; Kigezi District : Ishasha Gorge, Nov. 1946, *Purseglove* 2269 ! ; Masaka District : Bugala Is., Sept. 1950, *Philip* 385 !

TANGANYIKA. Bukoba District : Rubare, July 1931, *Eggeling* 6236 ! ; Lushoto District : W. Usambara Mts., Baga–Bumbuli road, 2·5 km. NE. of Sakarani, May 1953, *Drummond & Hemsley* 2405 ! ; Njombe, Aug. 1933, *Greenway* 3511 !

DISTR. **U**1, 2, 4 ; **T**1, 3, 4, 6–8 ; Belgian Congo, Nyasaland, Northern Rhodesia, ?Angola ; certain examples from Ghana and Ivory Coast appear to be this subspecies

HAB. Upland rain-forest (often dominant) and riverine forest in *Brachystegia* woodland : 1000–2100 m.

SYN. *P. holstii* Engl., P.O.A. C : 423, 191 (1895) ; Hauman & Balle in B.J.B.B. 21 : 191 (1951) ; I.T.U., ed. 2 : 333 (1952) ; F.C.B. 3 : 59 (1952) excl. syn. *P. tenuifolia*.

P. salicifolia Engl., P.O.A. C. : 191 (1895), *non* (Presl) Miq. (1855), *nom illegit.* Type : as *P. excelsa* subsp. *holstii*.

P. mildbraedii Engl. in Z.A.E. : 227, t. 23 (1911). Type : Ruanda-Urundi, W. Rugege Forest, *Engler* 1036 (B, holo.†)

P. excelsa Sabine var. *fulvescens* Engl. in Z.A.E. : 227 (1911). Types : Belgian Congo, between Beni and Irumu, *Mildbraed* 2808 (B, syn.†) & between Mawambi and Awakubu, *Mildbraed* 3232 (B, syn.†, BR, fragment, isosyn. !) & Spanish Guinea, Ikonangi, *Tessmann* 893 (B, syn.†)

[*P. excelsa* sensu T.T.C.L.: 476 (1949) ; I.T.U., ed. 2: 334, fig. 70d (1952) & sensu F.W.T.A., ed. 2, 1 : 429 (1958), pro parte ; *non* Sabine]

NOTE. *P. whytei* Engl. in E.J. 26 : 378 (1899) ; T.T.C.L.: 477 (1949). Type : Nyasaland, Mt. Malosa, *Whyte* (K, holo. !). This appears to be a form of subsp. *holstii* differing in having slightly larger flowers. Specimens seen from our area which have been so named are however normal subsp. *holstii*.

Subsp. *excelsa*, which replaces subsp. *holstii* in West Africa, differs in its larger flowers (6–8 mm. long), longer filaments (2·5–3 mm. long), staminodes (1·75–2 mm.) and styles (4–5 mm.), and in its indumentum which is silvery, not brown or fulvous.

3. **P. curatellifolia** [*Planch. ex*] *Benth.* in Hook., Niger Fl. 333 (1849) ; Hutch. in K.B. 1931 : 243 (1931) ; R. Grah. in K.B. 1957 : 229 (1957). Types : Senegal, *Heudelot* 362 (K, syn. !) & Nigeria, R. Quorra [R. Niger] at Patteh, *Vogel* 177 (K, syn. !)

A tree up to 15 m. tall but often considerably less, the bark corky, longitudinally fissured, and the crown rounded in well-grown examples. Leaves petiolate, variable in shape, but usually oblong or oblong-elliptic, 6·5–12 (–17) × 3·5–6(–9) cm., apically rounded (sometimes emarginate), basally cuneate to cordate (sometimes asymmetrically), coriaceous or ± so, green and glabrous above or with a tomentose midrib, silvery-grey to fulvous-brown-tomentose beneath, sometimes very thickly so, with very prominent midrib and close, subparallel primary nerves. Inflorescence an open or dense, terminal or axillary panicle, with abundant silvery to fulvous hairs ; cymes 2-3-flowered. Flowers 4–8 mm. long, sweetly smelling. Calyx rather asymmetrically cup-shaped, ± gibbous ventrally ; receptacle internally very shaggy at the throat ; calyx-lobes triangular, 1·75–4 mm. long (the dorsal pair a little longer than the others), acute, erect or spreading at anthesis. Petals white or very pale mauve, not exceeding the calyx-lobes. Filaments of fertile stamens 1·5–2·75 mm. long ; staminodes 0·5–1·5 mm. long, often ± hidden by the throat hairs. Ovary shaggy ; style 3–5 mm. long. Drupe ellipsoid-ovoid, up to 3·5 × 2·5 cm., edible, yellow when fresh, dark brown when dry and covered in variable density with light-brown patches of cork.

subsp. **curatellifolia**

Indumentum of inflorescence and of leaf-undersurfaces silvery or grey, that of the inflorescence appressed or slightly spreading. Flowers 4–6 (average 5·4) mm. long. Calyx-lobes 1·75–3·5 (average 2·8) mm. long.

UGANDA. Ankole District : near Mulema, *Bagshawe* 208 ! ; Masaka District : Katera Hill, *Maitland* 820 ! & S. Buddu, *Dawe* 324 !
KENYA. S. Kavirondo District : Kapogi, Oct. 1946, *Glasgow* 45/59 !
TANGANYIKA. Mwanza District : Geita district, Karumo, Chamabanda, Mar. 1953, *Tanner* 1246 ! ; Mpanda, Aug. 1954, *F. G. Smith* 1194 ! ; Uzaramo District : Mogo Forest Reserve, Aug. 1953, *Semsei* 1284 !
ZANZIBAR. Zanzibar Is., Walezo, Feb. 1930, *Vaughan* 1196 ! & Masingini Ridge, Feb. 1929, *Greenway* 1293 ! ; Pemba, Tasini, Dec. 1930, *Greenway* 2746 !
DISTR. **U**1, 2, 4 ; **K**5 ; **T**1, 4–7 ; **Z** ; **P** ; Senegal, Belgian Congo, Sudan and south to Tanganyika where it occurs with and is largely replaced by the next subspecies ; also here and there in South Africa and in the Seychelles.
HAB. Deciduous woodland, especially *Brachystegia* woodland extending to its upper limits, and then scattered in upland grassland ; often persisting in cultivated land and present in secondary bushland ; 0–2070 m.

SYN. *P. curatellifolia* Benth., *sensu stricto* ; F.T.A. 2 : 368 (1871) ; I.T.U., ed. 2, 333, fig. 70 a–c (1951) ; F.C.B. 3 : 66 (1952)

subsp. **mobola** (*Oliv.*) *R. Grah.* in K.B. 1957 : 229 (1957). Type : Angola, Huilla, *Welwitsch* 1282 (K, lecto. !, BM, isolecto. !)

Indumentum of inflorescence and leaf-undersurfaces rather thicker than in the preceding and fulvous-brown, that of the inflorescence more spreading. Flowers rather longer, 5–8 (average 6·15) mm. Calyx-lobes 2–4 (average 2·95) mm. long.

KENYA. Kwale District : 9·6 km. from the Kwale–Tanga road end of the Mrima road, Sept. 1957, *Verdcourt* 1932 !
TANGANYIKA. Buha District : 48 km. S. of Kibondo on Kasulu road, July 1951, *Eggeling* 6220 ! ; Iringa District : near Ipana, Aug. 1933, *Greenway* 3599 ! ; Njombe District : near Lupembe, N. of the Upper Ruhudje R., Aug. 1931, *Schlieben* 1159A !
DISTR. **K**7 ; **T**4, 7 ; **P** ; the Sudan, the Belgian Congo and south to South Africa ; more southern in its range than the preceding subspecies.
HAB. As for subsp. *curatellifolia* ; 0–1800 m.

SYN. *P. mobola* Oliv. in F.T.A. 2 : 368 (1871) ; T.T.C.L. : 476 (1949) ; Pardy in Rhod. Agric. Journ. 48 : 264 (1951) ; F.C.B. 3 : 64 (1952) ; Brenan in Mem. N. Y. Bot. Gard. 8 : 430 (1954)
[*P. curatellifolia* sensu auct. plur., *non* Benth.]

NOTE. This and the preceding subspecies are not always easily differentiable owing to intermediate characters.

4. **P. goetzeniana** *Engl.* in E.J. 34 : 153 (1904) ; T.T.C.L. 476 (1949). Type : Tanganyika, E. Usambara Mts., *Engler* 496 (B, holo.†)

An evergreen tree, 25–50 m. high, with reddish-brown bark and a ± spreading rounded crown. Leaves coriaceous, elliptic to oblong-elliptic, 8–15 × 4–6·5 cm., acuminate, basally broadly cuneate, brownish-green, glabrous and somewhat shining above, matt beneath ; primary nerves 8–10. Inflorescence a much branched broadly corymbose greyish panicle up to 12 cm. broad, seldom exceeding 8 cm. long ; young branchlets and pedicels

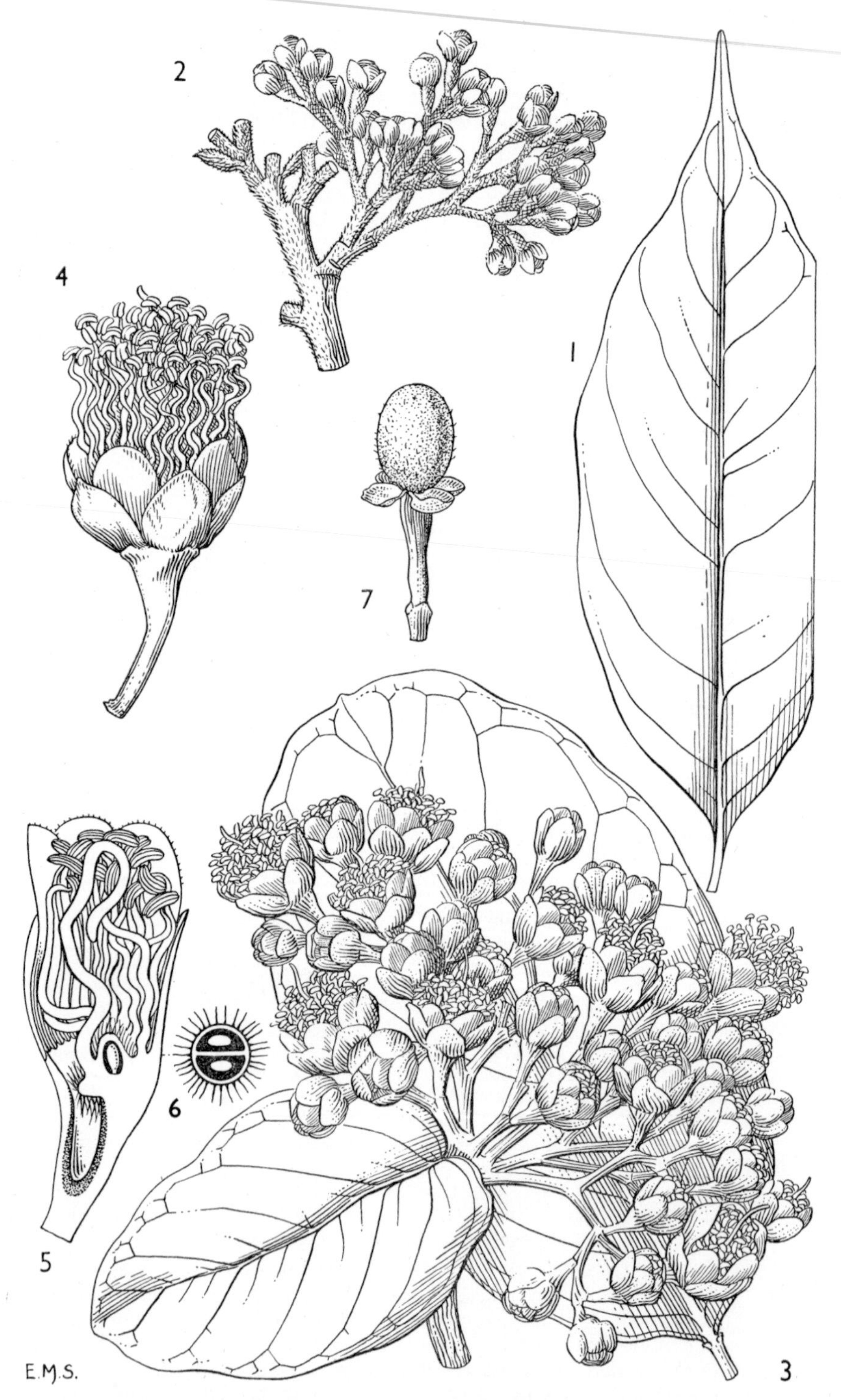

FIG. 7. *PARINARI GOETZENIANA*—**1,** leaf, × 1; **2,** part of the inflorescence, × 1; *PARINARI POLYANDRA* subsp. *FLORIBUNDA*—**3,** leaf and inflorescence, × 1; **4,** flower, × 2; **5,** flower in L.S. to show stamens and pistil, × 3; **6,** ovary in T.S., diagrammatic; **7,** immature fruit, × 1. 1, 2, from *Greenway* 1588; 3–6, from *Eggeling* 6088; 7, from *Hornby* 2002.

covered with thick grey or greyish-brown appressed or erect-patent hairs. Flowers white or greenish-white, 9–11 mm. long (from the joint in the pedicel to the tips of the sepals). Calyx turbinate, fleshy, tomentose, not or scarcely narrowed at the base, the lobes 4–5·5 mm. long, rounded or very broadly oblong. Petals broadly oblong, concave, 7–8(?–10) mm. long. Fertile stamens many, the filaments tortuous, ± 15 mm. long ; staminodes very short, up to 1 mm. long, sometimes scarcely discernible. Ovary shaggy with long yellowish hairs ; style ± 15 mm. long, fleshy. Drupe edible, black when ripe. Fig. 7/1, 2.

Tanganyika. Lushoto District : E. Usambara Mts., Amani, Sept. 1929, *Greenway* 1728 ! & May 1935, *Greenway* 4044 ! ; Amani–Kiumba road, Aug. 1913, *Grote* 6914 !
Distr. **T3** ; apparently confined to the E. Usambara Mts.
Hab. Evergreen rain-forest ; about 900 m.

Note. This species clearly stands close to the West African *P. glabra* Oliv. and to *P. robusta* Oliv. From the former it is distinguished largely by having a thick indumentum on the inflorescence (although the inflorescence branches of the type specimens of *P. glabra* are shortly but undeniably pubescent, contrary to its type description). From *P. robusta* our species is distinguishable by its receptable being rather longer than broad, the reverse characterizing *P. robusta*. Further collecting of *P. goetzeniana* will indicate whether these three species should be treated in subspecific relationship.

5. **P. polyandra** *Benth.* in Hook., Niger Fl. 333 (1849) ; R. Grah. in K.B. 1957 : 229 (1957). Type : Nigeria, R. Quorra [R. Niger] at Attah, *Vogel* (K, holo. !)

A tree up to about 12 m. tall, sometimes low and bushy, with finely-fissured, ± smooth bark. Leaves shortly petiolate, broadly ovate or ovate-elliptic, 8–11·5(–15) × 4–6(–7·5) cm., glabrous, shining and green or brownish-green above, matt and green and varying from glabrous to grey-tomentose beneath, rounded (sometimes emarginate) or more rarely obtuse, basally broadly cuneate to subcordate ; nerves usually clearly visible on both sides, the primaries numbering 8–9 " pairs." Inflorescence terminal, corymbose, broader than long, 8(–10) cm. long, 13(–18) cm. broad, borne on ± thick branchlets varying from shaggily hairy to glabrous. Flowers fleshy, about 1·3–2 cm. long (excl. stamens). Calyx turbinate, enlargening upwards but slightly asymmetrical, 7–15 mm. long (measured from the joint in the pedicel to the base of the sepals), glabrous to thickly shining-tomentose, the lobes ± rounded, concave, 4·5–8 mm. long. Fertile stamens 40–50(–60) ; staminodes very short, usually reduced to minute prominences on the fleshy receptacle-rim. Ovary shaggy ; style 13–22 mm. long. Drupe oblong-ovoid, rather asymmetrical, glabrous to tomentellous, up to 3 × 2·2 cm.

subsp. **floribunda** (*Bak.f.*) *R. Grah.* in K.B. 1957 : 230 (1957). Type : Nyasaland, Karonga District, Fort Hill, *Whyte* (K, holo. !)

Calyx 10–15 mm. long (measured as stated above) ; sepals 5–8 mm. long. Style ± 22 mm. long. Fig. 7/3–7.

Tanganyika. Buha District : between Kanyato and Makere, July 1951, *Eggeling* 6216 ! ; Mpanda District : Kabungu Forest Reserve, Aug. 1948, *Semsei* 2583 ! ; Chunya District : Kipembawe, May 1951, *Eggeling* 6088 !
Distr. **T4, 7** ; Belgian Congo, Northern and Southern Rhodesia, Nyasaland.
Hab. Upland bushland, woodland and scattered tree grassland ; 1000–1700 m.

Syn. *P. floribunda* Bak.f. in K.B. 1897 : 265 (1897)
P. bequaertii De Wild. in F.R. 13 : 108 (1914) ; F.C.B. 3 : 56 (1952). Type : Belgian Congo, Katanga, Elisabethville, *Bequaert* (BR, holo.)
?*P. bequaertii* De Wild.; T.T.C.L. : 476 (1949)

Note. Subsp. *polyandra*, which occurs in West Africa, extending eastwards to Sudan, differs in having a shorter calyx (7–10 mm. long), shorter sepals (4·5–5 mm. long), and a shorter style (about 13 mm. long).

9. HIRTELLA

L., Sp. Pl. : 34 (1753) & Gen. Pl., ed. 5 : 20 (1754)

Trees with hermaphrodite flowers. Leaves simple, entire. Calyx-tube narrowly cylindrical, usually gibbous, with 5 spreading or reflexing lobes ; receptacle-mouth nearly circular, crateriform. Petals 5, caducous. Antheriferous stamens 7–9, long-exserted, free to the base, inserted on the dorsal rim of the receptacle mouth ; staminodes very short, free to the base, inserted on the rim ventrally. Ovary unilocular, inserted dorsally but appearing ± central in the receptacle mouth ; ovules basal, 2 (apparently only 1 usually developing) ; style basal, filiform, long-exserted. Fruit with a hard crustaceous pericarp.

Largely a South American genus but with two endemic species in Africa.

Fruit not exceeding 2·6 × 1·2 cm. ; leaves with obscurely reticulate venation below, roughly twice as long as broad . . . . 1. *H. zanzibarica*
Fruit 3·5–4 × 1·8 cm.; leaves with markedly reticulate venation below, roughly three times as long as broad 2. *H. megacarpa*

1. **H. zanzibarica** *Oliv.* in Hook., Ic. Pl. 12 : 81, t. 1193 (1876) ; Brenan in Trop. Woods No. 86 : 5 (1946) ; T.T.C.L. : 475 (1949). Type : Tanganyika, Mafia Island, *Kirk* (K, holo. !)

An evergreen tree up to 25 m. tall with blackish bark, or rarely a shrub. Young branchlets closely pubescent to densely villous with spreading light brown hairs ; older branches commonly glabrescent, black or almost so. Leaves shortly petiolate, leathery, shining (especially on the upper surface), ovate-elliptic to oblong-elliptic, 3·8–8·5 × 1·4–4·3 cm. (commonly about 8 × 4 cm.), ± acute or shortly acuminate, basally short-cuneate to subcordate but commonly rounded or ± so, glabrous (or pubescent on the midrib and nerves beneath when immature), reticulately nerved on both sides, but more obscurely so beneath. Inflorescence a terminal or axillary dense panicle up to 15(–18) × 11(–14) cm., the branchlets velvety ; lower bracts ovate to oblong-lanceolate, ± densely pubescent and eglandular ; upper bracts ovate-reniform, usually with many obvious unequally-stalked glands on the margins (rarely eglandular, or with the glands sessile and hidden in the indumentum). Flowers greenish-white. Calyx-tube 6–9 mm. long, 1·1–1·8 mm. broad, slightly expanded above, abruptly enlarged at the base, hairy ; lobes oblong, 4–5 mm. long, usually glandular on the margins. Petals scarcely exceeding the calyx-lobes, oblong-elliptic to obovate, ± twisted. Stamens up to 1 cm. long (? longer), long-exserted ; staminodes very short. Ovary densely shaggy ; style 1·2 cm. long (? longer), with many ± spreading hairs in the lower part. Fruit dark brown, ellipsoid, basally narrowed, 1·8–2·3(–2·6) × 1·1–1·2 cm., obtuse ; pericarp externally subglabrous, densely lanate within. Fig. 8/5, 6, p. 57.

KENYA. Kwale District : Shimba Hills, Mwele Mdoga Forest, Feb. 1953, *Drummond & Hemsley* 1172 ! & Shimba Hills, May 1930, *Donald 22 in F.D.* 2367 ! ; Buda Forest, Nov. 1936 , *Dale* 3579 !

TANGANYIKA. Uzaramo District : Vikindu Forest Reserve, Aug. 1953, *Semsei* 1316 ! ; Rufiji District : Mafia Island, Tondwa, Oct. 1937, *Greenway* 5376 ! & Kilindoni, Aug. 1937, *Greenway* 4996 !

ZANZIBAR. Pemba, Kichange, *Vaughan* 377 ! & Chake Chake, Sept. 1929, *Vaughan* 652 !

DISTR. **K7** ; **T6**, 7 ; **P** ; Portuguese East Africa, Nyasaland & Madagascar.
HAB. Lowland rain-forest ; 0–900 m.

SYN. *Acioa goetzeana* Engl. in E.J. 30 : 315, t. 12 (1901). Type : Tanganyika, base of Livingstone Mts., Ikombe, *Goetze* 1176 (B, holo. †, BR, iso. !)
H. zanzibarica Oliv. var. *cryptadenia* Brenan in Trop. Woods No. 86 : 11 (1946). Type : Zanzibar, Pemba, Chake Chake, *Vaughan* 652 !

NOTE. Var. *cryptadenia* is separated from typical material on vegetative characters, which are, however, embraced by the general plasticity of the species. Its poor glandular development can be matched with material from Portuguese East Africa (*Dawe* 401 !) though in this specimen the inflorescence-indumentum is more normal. The more coppery-coloured leaves and apparently pendulous inflorescences of the variety do not seem to be other than minor variational characters. Thus, although var. *cryptadenia* may stand apart as an extreme form, its separation as a taxon is perhaps not fully justifiable.

2. **H. megacarpa** *R. Grah.* in K.B. 1957 : 231 (1957). Type : Tanganyika, W. Usambara Mts., Shagai Forest, *Drummond & Hemsley* 2614 (K, holo. !, EA, iso. !)

A tall tree, probably evergreen, up to 25 m. high. Branchlets pubescent. Leaves coriaceous, oblong-elliptic, up to 8 × 2·8 cm., acuminate to caudate-acuminate, basally cuneate or rounded, matt on both faces or somewhat shining above, glabrous except the midrib beneath which is pubescent with rather long hairs, clearly reticulately nerved but especially markedly so beneath ; the midrib of the under-surface very prominent but the primary nerves rather less so than in the preceding species. Inflorescences not seen. Flowers imperfectly known. Calyx-tube 11 m. long, apparently neither unilaterally gibbous nor abruptly enlarged at the base, slightly expanded above, deeply grooved in the upper part, thickly velutinous ; the lobes eglandular, ± 5 mm. long. Stamens 8 mm. (? or more) long, orientated as in the preceding species, and the staminodes likewise. Fruit oblong-ovoid, obtuse, basally narrowed, up to 3·5 × 1·8 cm. ; pericarp green, minutely pubescent outside, lanate within. Seed 1, red, with scattered surface hairs.

TANGANYIKA. Lushoto District : W. Usambara Mts., Shagai Forest, near Sunga, May 1953, *Drummond & Hemsley* 2614 ! ; Shagai Forest, May 1953, *Procter* 221 ! ; Iringa District : Nyumbanitu, Sept. 1958, *Ede* 68*
DISTR. **T3**, 7* ; not known elsewhere.
HAB. Upland rain-forest ; 1350–1950 m.

NOTE. Although the floral parts suggest some distinctions from *H. zanzibarica*, it must be recorded that only one imperfect flower was available for investigation. For the time being, therefore, this species is distinguished from *H. zanzibarica* on its vegetative and fruit characters, and the floral characters described above should be accepted with reserve.

10. MAGNISTIPULA

Engl. in E. J. 36 : 226 (1905)

Trees or shrubs with hermaphrodite flowers. Leaves simple, entire. Calyx-tube obliquely campanulate, with 5 erect or reflexed lobes, gibbous ; receptacle-mouth asymmetrically crateriform. Petals 5. Antheriferous stamens included, arcuate, attached to the dorsal rim of the receptacle-mouth ; staminodes inserted on the ventral rim, connate towards the base and disposed (in Tropical East African spp.) as a comb or as an elevated tongue. Ovary with an apical tuft of hairs, unilocular ; ovules basal, 2 (apparently only 1 usually developing) ; style basal, arcuate or ± so, not filiform, neither exserted nor exceeding the stamens, with a tuft of hairs at the base (in Tropical East African spp.). Fruit drupaceous, 1-seeded.

A genus formerly united with *Hirtella* but showing markedly different floral characters in which there is a greater affinity to *Parinari* (subgen. *Sarcostegia*). Confined to Africa and specifically most abundant in Belgian Congo.

* This Iringa District record is added by Editor in proof.

Leaves basally rounded to cordate, apically rounded (sometimes emarginate) to broadly acute (if acuminate then only very shortly so); inflorescence branches ascending at an angle of 45°–72° (–90°); largely a woodland tree, seldom exceeding 18 m. in height . . . . . 1 *M. bangweolensis*

Leaves narrowed to the base, apically obtuse to acutely acuminate; inflorescence divaricately branched, the branches spreading horizontally; rain-forest tree up to about 30 m. in height . 2. *M. butayei* var. *greenwayii*

1. **M. bangweolensis** (*R. E. Fries*) *R. Grah.* in K.B. 1957 : 230 (1957). Types : Northern Rhodesia, near Lake Bangweulu, Kawendimusi, *Fries* 780 (S, syn.) & Kamindas, *Fries* 780A (S, syn.) & R. Mano, *Fries* 732 (S, syn., K, isosyn. !)

A large tree up to about 18 m. tall (sometimes more), with grey-brown coarsely reticulate bark and a dense rounded crown. Young branches velutinous with olive-golden hairs, later glabrescent and grey. Leaves broadly oblong-elliptic to obovate, up to 9·5–15 × 4·5–8·7 cm., rounded (sometimes emarginate) to shortly and obtusely acuminate, basally cordate, shining, brown and glabrous above except for the pubescent midrib, matt, yellow-green and pubescent beneath except when fully mature and then glabrous on the lamina; primary nerves very prominent beneath. Inflorescence paniculate, terminal or axillary, the racemes 4–6 cm. long, ± ascending; bracts linear-lanceolate, with the pedicels and calyx densely velutinous with olive-golden hairs. Calyx-tube ± 5 mm. long; the lobes ovate-deltoid to lanceolate. Petals white, oblong-elliptic, 5–5·5 mm. long. Fertile stamens 7–8, about 4 mm. long; filaments ± unevenly connate towards the base; anthers yellow; staminodes (1·0–)1·5–2·0 mm. long, irregularly connate in the lower half. Ovary sessile or nearly so, inserted dorsally near the receptacle-rim. Drupe velutinous, elliptic, 3·5 × 2 cm., very obtuse, slightly oblique-sided, olive-black. Fig. 8/1–4.

TANGANYIKA. Buha District : Nyamgalika, Aug. 1950, *Bullock* 3199 ! ; Mpanda District : Kabungu, July 1948, *Semsei* 120 ! ; Mbeya District : Mbozi, Sept. 1932, *Jessel* 34 !

DISTR. **T4**, **7** ; Northern Rhodesia and Nyasaland.

HAB. Deciduous woodland and wooded grassland, perhaps especially in sandy soils, fixed sand-hills, and on termite-mounds ; 480–1500 m.

SYN. *Parinari bangweolensis* R. E. Fries in F.R. 12 : 540 (1913)
Hirtella bangweolensis (R. E. Fries) Greenway in K.B. 1928 : 199 (1928) ; F.C.B. 3 : 40 (1952)

NOTE. It is possible that the relationship of *M. bangweolensis* to *M. butayei* De Wild. is closer than has been supposed. The characters for distinction are apparently not wholly reliable though the leaves of the latter (though not in its var. *greenwayii*) appear always to have a glabrous midrib on the undersurface. The differences in the number of staminodes, as given in various works, and in other flower characters do not hold strongly, and it is doubtful whether any such character can be reliably used for purposes of distinction. Apart from rather extreme material (which gives adequate distinction in leaf-shape), there is reason to suggest that both species should be more properly regarded in subspecific relationship to each other, but material to hand is as yet hardly adequate for a decisive treatment. *Bullock* 3253 (from Tanganyika, Kigoma District, 65 km. S. of Uvinza) has narrower leaves than the bulk of material of *M. bangweolensis* that has been examined, and its panicle is rather more open than is apparently normal for this species, but the leaves are pubescent below (though largely on the midrib), and on this character it is here left as *M. bangweolensis*. The genus requires investigation on a monographic basis.

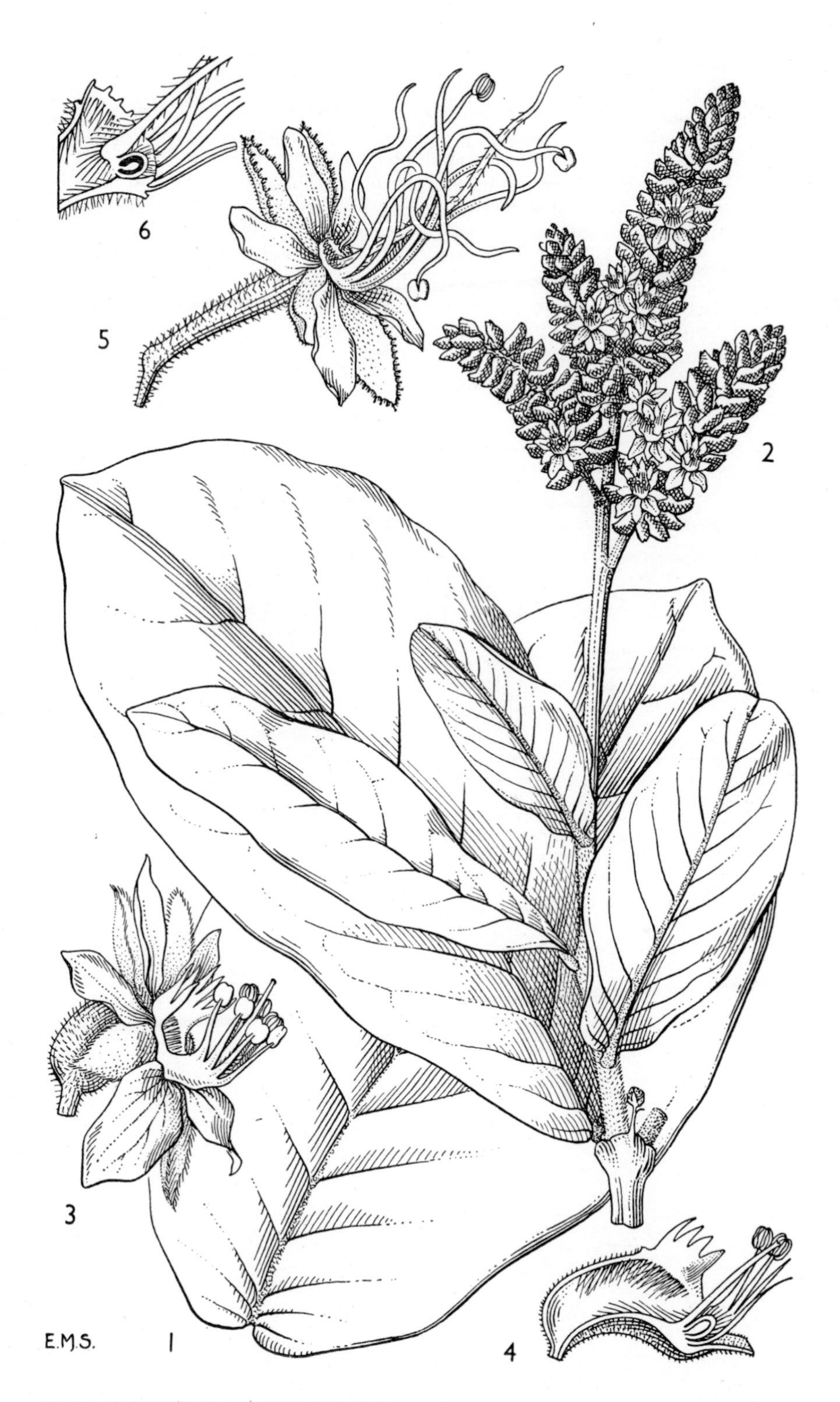

FIG. 8. *MAGNISTIPULA BANGWEOLENSIS*—**1,** leaf, × 1; **2,** flowering branch, × 1; **3,** flower, × 4; **4,** flower, in L.S. to show position of ovary, × 4; *HIRTELLA ZANZIBARICA*—**5,** flower, × 4; **6,** part of flower in L.S. to show position of ovary, × 6. **1,** from *Bullock* 3199; 2–4, from *F. G. Smith* 1199; 5–6 from *Gomes e Sousa* 781.

2. **M. butayei** *De Wild.* in Ann. Mus. Congo Belg., Sér. 5, 2 : 255 (1908). Type : Belgian Congo, basin of R. Nsele, *Gillet in Butaye* (BR, holo.)

An evergreen tree up to 30 m. tall, with a fairly smooth grey bark and a long-branched columnar crown. Young branches velutinous with rusty-brown hairs, later glabrescent and grey. Leaves oblong-elliptic to oblong-lanceolate, 8·5–15·5 × 3·2 × 5·0 cm., attenuate to each end, or rounded to cordate at the base, acutely acuminate, shining on both faces, glabrous except for the midrib and primary nerves above, and glabrous beneath or with rather long rusty hairs on the very prominent midrib and primary nerves. Inflorescence paniculate, axillary (? also terminal), the branches ascending or arising ± at right angles. Flowers white, ± 7 mm. long, Calyx-tube 4 mm. long ; lobes triangular, ± acute. Filaments 3–4 mm. long, connate towards the base ; staminodes 1.5–1·75 mm. long but variable. Ovary sessile or nearly so, inserted near the rim of the receptacle.

SYN. *Hirtella butayei* (De Wild.) Brenan in Trop. Woods No. 86 : 4 (1946) ; F. C. B. 3 : 41 (1952)

var. **greenwayii** (*Brenan*) *R. Grah.* in K. B. 1957 : 406 (1958). Type : Tanganyika, E. Usambara Mts., Amani, *Greenway* 6161 (K, holo. !, EA, FHO, iso. !)

Leaves narrowed to the base ; midrib beneath hairy. Indumentum of the inflorescence rather coarse, spreading, pale rusty-brown.

TANGANYIKA. Lushoto District : Amani, Apr. 1941, *Greenway* 6161 ! & Apr. 1922, *Soleman* G6133 !
DISTR. **T3** ; also in the Belgian Congo (Haut-Katanga and Bas-Katanga)
HAB. Rain-forest, steep slopes (but information scanty) ; 840 m.

SYN. *Hirtella sapinii* (De Wild.) A. Chev. var. *greenwayii* Brenan in Trop. Woods No. 86 : 4 (1946) ; T.T.C.L. : 475 (1949)
Hirtella butayei (De Wild.) Brenan var. *greenwayii* (Brenan) Hauman in B.J.B.B. 21 : 179 (1951) ; F.C.B. 3 : 41 (1952)

NOTE. I follow Hauman in classifying var. *greenwayii* in close kinship with *M. butayei* rather than with the larger-flowered *M. sapinii* De Wild., but in view of the preceding comments (under *M. bangweolensis*) the possibility of a more correct affinity to this latter species should not be ignored.

Var. *butayei*, which is found in the Belgian Congo but not, so far recorded from our area, differs in having a glabrous midrib on the undersurface of the leaves which are usually rounded to cordate at the base, and in the inflorescence indumentum which is largely appressed and fuscous.

INDEX TO ROSACEAE